BEYOND DROUGHT
PEOPLE, POLICY AND PERSPECTIVES

The effects of major drought

1864–66　　All States affected except Tasmania.

1880–86　　Southern and eastern States affected.

1895–1903　Sheep numbers halved and more than 40 per cent loss of cattle. Most devastating drought in terms of stock losses.

1911–16　　Loss of 19 million sheep and two million cattle.

1918–20　　Only parts of Western Australia free from drought.

1939–45　　Loss of nearly 30 million sheep between 1942 and 1945.

1963–68　　Widespread drought. Also longest drought in arid central Australia: 1958–67. The last two years saw a 40 per cent drop in wheat harvest, a loss of 20 million sheep, and a decrease in farm income of $300–500 million

1972–73　　Mainly in eastern Australia.

1982–83　　Total loss estimated in excess of $3000 million. Most intense drought in terms of vast areas affected.

1991–95　　Average production by rural industries fell about 10 per cent; $590 million drought relief provided by the Commonwealth Government.

2002–03　　Economic growth in Australia reduced by about $7 billion. The Australian Government has committed more than $1 billion in assistance to farm families.

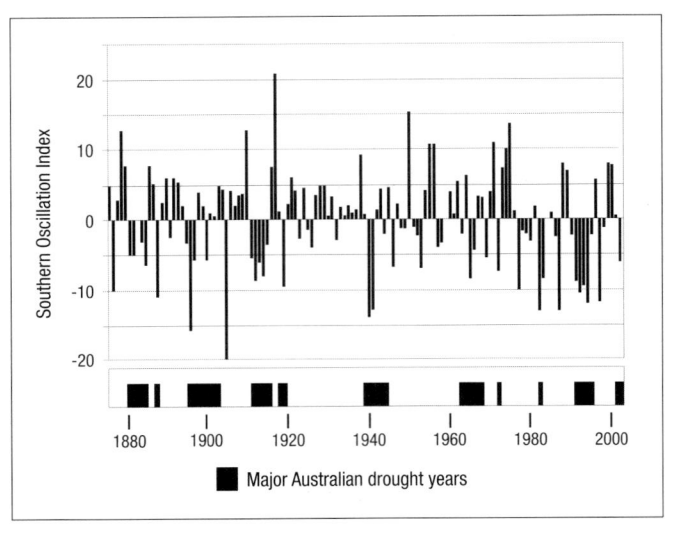

Source: Australian Bureau of Meteorology, http://www.bom.gov.au/lam/climate/levelthree/cpeople/drought.htm

BEYOND DROUGHT
PEOPLE, POLICY AND PERSPECTIVES

Linda Courtenay Botterill and Melanie Fisher (Editors)

CSIRO
PUBLISHING

National Library of Australia Cataloguing-in-Publication entry
Beyond drought: people, policy and perspectives.

Bibliography.
ISBN 0 643 06954 2.
        1. Droughts – Australia. 2. Droughts – Government policy –
        Australia. I. Botterill, Linda Courtenay. II. Fisher, Melanie.
338.180994

Available from:
CSIRO PUBLISHING
150 Oxford Street (PO Box 1139)
Collingwood Vic. 3066
Australia

Telephone: +61 3 9662 7666
Freecall: 1800 645 051 (Australia only)
Fax: +61 3 9662 7555
Email: publishing.sales@csiro.au
Website: www.publish.csiro.au

Front cover
Drought near Hillston, NSW by Gregory Heath, courtesy CSIRO Land and Water

Set in Stempel Schneidler 10/12.5
Cover and text design by James Kelly
Typeset by James Kelly
Printed in Australia by Ligare

The views expressed in this work are each author's own and do not necessarily reflect those of their employers, referees or editors.

Contents

Foreword and acknowledgements

When the idea of this book was conceived, Australia's rural sector was buoyant. The 2001–02 winter crop was near record levels and beef prices were healthy. Adequate rain had fallen in a timely fashion for a number of seasons and drought was an unpleasant memory from the mid-1990s. We thought that this environment provided the perfect opportunity to contribute to a rational, considered discussion of drought in Australia—its social, environmental and economic impacts, and the appropriate responses from government, farmers and the broader community. As it turned out an El Niño visited our shores in 2002 and we found ourselves writing about drought as it was occurring. This necessarily put pressure on many of our authors whose professional lives are tied up with drought and who therefore found themselves very busy at work while being hassled by us to meet, admittedly tight, deadlines. We would therefore like to thank our authors up front for putting up with our nagging and delivering what we believe are very valuable contributions to this book. In early January, as we were working on this manuscript, Canberra experienced devastating bushfires, exacerbated no doubt by the widespread drought. This highlighted for many of us that drought not only affects farmers, but our homes in the cities as well.

As people with an interest in rural policy in particular and public policy more generally, we thought it would be valuable to bring together a range of experts with different disciplinary perspectives to discuss drought and what it means for Australia. Australia really is a land of drought and flooding rains, and we were concerned that the general public and the media are not as well informed about the impact of climate variability as we are about interest rates and budget deficits. This book sets out to redress some of that gap.

We did not want this book to be a collection of stand-alone essays, interesting in their own right but lacking cohesion, so we set in train a process that involved two workshops of our authors. They discussed the overall direction of the book and identified themes early on that would be reflected in all our work. We would like to thank Patrick Troy from the Centre for Resource and Environmental Studies at the Australian National University for his valuable advice early in the book's development on how to manage the editing process.

To our knowledge, the last time an interdisciplinary book on drought was produced in Australia, it was edited by J.V. Lovett in 1973.[1] In a nice piece of symmetry, Professor Lovett is now the Managing Director of the Grains Research and Development Corporation, the sponsors of our authors' workshops. We are very grateful for the interest Professor Lovett showed in the concept of the book from the outset and for the financial support given to allow us to run our workshops. The first workshop focussed on identifying the key themes to be explored in the book. These were suggested early in the day and proved surprisingly robust in the face of enthusiastic discussion. The second workshop had a more serious purpose. In addition to our authors, it was attended by expert discussants who were given copies of the draft chapters. They then led the discussion of each chapter and their contributions were invaluable. We would like to thank Professor Snow Barlow, David Dumaresq, Dr Frank Lewins, Professor John Quiggin, Associate Professor Ian Ward, Dr David White and Dr Barry White for participating in the workshop and providing us with the benefit of their knowledge and experience. We greatly appreciated the assistance of Debbie Dewey, whose energy and organisational ability was central to the smooth running of the workshops, and whose deft organisation of Melanie's diary allowed for her ongoing involvement in the project. Dr Greg Laughlin, Dr Bob Munro and Penny Carter all provided advice that was greatly appreciated, as was the feedback from the many of our colleagues who read the draft manuscript.

Also important to the production of this book was the sponsorship from the Rural Industries Research and Development Corporation—our thanks to Dr Simon Hearn for his support for this project.

Particular thanks are due to Professor Elim Papadakis, Director of the National Europe Centre, and Dr Peter O'Brien, Executive Director of the Bureau of Rural Sciences, for their support of our continuing work on this project well into 2003.

We would also like to thank the men in our lives, Bob, Tim and Max for their ongoing patience, love and support.

This book is intended to place drought on the public agenda as a topic of considerable importance to all Australians. Responding in a timely and equitable manner to the needs of farm businesses affected by inadequate rainfall is a public policy challenge. We do not provide any single solution, rather we have set out to highlight different perspectives on drought in the hope of stimulating public debate and contributing to better drought policy.

And this raises our final point—we would like to see the word 'drought' removed from the national lexicon. It is an inheritance from our European history, which has its roots in a predictable, annual climate cycle. As we learn more about our unique island continent we become more aware that this is not a particularly accurate or constructive way of viewing the world. We would like to see discussion of 'climate variability' policy in recognition that there are not only El Niño events but La Niñas as well, and both are a natural part of the Australian climatic cycle. We would like to see policy move away from a crisis response to a set of policy instruments that meets the needs of farm families, their communities, the environment and the broader economy in a way that is in harmony with Australian biophysical and climate reality.

Linda Courtenay Botterill and Melanie Fisher
Canberra
May 2003

---

1 Lovett, 1973

Contributors

Linda Courtenay Botterill is a Postdoctoral Fellow in the National Europe Centre at the Australian National University. She has extensive experience in public policy, having worked in the Commonwealth Public Service, as a ministerial adviser and as a policy officer in two industry associations before undertaking her PhD in political science at the ANU. In 1993 and 1994, she advised the Commonwealth Minister for Primary Industries and Energy on drought and rural adjustment. Her research interest is the rural policy development process, and her area of expertise is structural adjustment in the rural sector and drought policy.

Peter Cox has a PhD in Technological Economics from Stirling University in Scotland. He has worked in several developing countries in Africa and south-east Asia. He spent eight years with CSIRO (at Narrabri with the Division of Plant Industry, and Toowoomba with the Division of Tropical Crops and Pastures) after moving to Australia from Papua New Guinea in 1988. More recently, he worked with the International Crops Research Institute for the Semi-arid Tropics (ICRISAT) in India and the International Rice Research Institute (IRRI) in Cambodia as part of their social science programs. At present, he is Regional Technical Adviser for Agriculture and Natural Research Management for south-east Asia with Catholic Relief Services, a US-based NGO.

Melanie Fisher is Deputy Executive Director of the Bureau of Rural Sciences in the Australian Government Department of Agriculture, Fisheries and Forestry. Her areas of responsibility in the Bureau include climate and agriculture, forests and vegetation, and social sciences. She has wide experience in agricultural policy and in 1995 was the drought adviser to the then Commonwealth Minister of Primary Industries and Energy. Her areas of interest include risk perception and communication, public policy processes and the role of science in policy development. She has graduate training in psychology and a Masters in Public Policy from the ANU.

Peter Hayman leads the NSW Agriculture agroclimatology unit, which was launched in 1999 to provide a better link between climate science and agricultural systems in NSW. He is a member of the World Meteorological Organisation expert team on end-user liaison. After completing a masters degree in crop physiology, Peter Hayman worked as an extension officer before completing a PhD in agroclimatic risk management at the University of Western Sydney. He has received research grants to work with grain farmers on their management of climate risk in the north-eastern and southern grain belt, and with wool producers in the NSW rangelands.

Janette Lindesay has a PhD in Climatology from the University of the Witwatersrand in South Africa, where she led the Climatology Research Group (CRG) in research into southern African climate variability and drought in the early 1990s. The CRG developed prototype seasonal climate outlooks for southern Africa during that time, with an active program of extension to primary producers and agriculturally-based industry. She moved to Australia in 1993 to take up an appointment as Senior Lecturer in Climatology at the ANU, where she is engaged in research on low-frequency variability in the El Niño Southern Oscillation and its impacts, and on changing rainfall seasonality in Australia. She has co-authored a number of papers and an authoritative text on El Niño–Southern Oscillation Cycle (ENSO) and climate variability, and has co-authored or edited three other books. Her other research interests include climate change, climate impacts, and climatological aspects of wildfire in the tropics and subtropics. At present she is Education Manager in the Cooperative Research Centre for Greenhouse Accounting, based at the ANU.

Bruce O'Meagher is a senior Commonwealth Public Servant. He has worked in several departments, including the Treasury and the Departments of Primary Industries and Energy and Industry, Tourism and Resources. He has worked on rural and drought-related policy development, and has authored and co-authored a number of papers on drought policy development.

Mark Stafford Smith has worked at CSIRO's Centre for Arid Zone Research in Alice Springs for two decades, with an emphasis on management responses to climatic variability in rangelands grazing enterprises for most of that time. He has had several encounters with the design, implementation and implications of drought policy, from input to the National Drought Policy, through surveys and modelling of pastoralists' responses to drought in different rangeland regions, and involvement in regional adjustment committees, to on-farm analyses of the effects of taxation and other policy instruments on pastoralist decision-making.

Daniela Stehlik is Associate Dean Research in the Faculty of Arts, Health and Sciences, and Associate Professor in Sociology in the School of Psychology and Sociology at Central Queensland University. From December 2003, she will be the inaugural Professor of Stronger Communities at Curtin University in Perth. Her research focuses on the intersections of community resiliency, human service practice and social cohesion in regional/rural Australia. Her specific interests are in ageing, disability, gender, power and community development. She has published widely in Australia and internationally, and regularly presents her work at national and international conferences. She is a member of the *Rural Society* editorial board, undertakes reviews for national journals and is The Australian Sociological Association's Jean Martin Award Convenor.

Åsa Wahlquist is the rural writer for *The Australian*, a position she has held since 1997. She has a degree in Agricultural Science from the University of Adelaide. After finishing university she worked on the family property, Botobolar Vineyard, at Mudgee in NSW. Åsa has worked as a rural journalist at the ABC, on *Country Hour* on ABC Radio and *Countrywide* on ABC TV, as well as in print. She has received a number of awards, including a Walkley Award in 1996 for a three-part series published in *The Land*, entitled 'The Gutting of Rural NSW', and the 1993 European Community Journalist Award.

Donald Wilhite is Director of the National Drought Mitigation Center and the International Drought Information Center, and Professor, School of Natural Resources, University of Nebraska-

Lincoln. Dr Wilhite's research and outreach activities focus on issues of drought monitoring, planning, and mitigation. He has collaborated with numerous countries and regional and international organisations on drought policy and planning issues. He has authored or co-authored more than 100 journal articles, monographs, book chapters, and technical reports. He is editor of several books, including *Drought: A Global Assessment*, published in 2000 by Routledge Publishers, as part of a seven-volume series on natural hazards and disasters.

Abbreviations

ABC	Australian Broadcasting Corporation
AUSLIG	Australian Surveying and Land Information Group (now part of Geoscience Australia)
BSE	Bovine Spongiform Encephalopathy ('Mad Cow' Disease)
CAS	Complex adaptive systems
CO_2	Carbon dioxide
CQ	Central Queensland
CWA	Country Women's Association
ENSO	El Niño Southern Oscillation Cycle
FMDs	Farm Management Deposits
LUCNA	Land Use Change in Northern Australia
NDMC	National Drought Mitigation Center (University of Nebraska, Lincoln)
NDP	National Drought Policy
NDPC	National Drought Policy Commission (USA)
NGOs	Non-governmental organisations
NSW	New South Wales
R&D	Research and development
RAS	Rural Adjustment Scheme
RASAC	Rural Adjustment Scheme Advisory Council
RIRDC	Rural Industries Research and Development Corporation
SOI	Southern Oscillation Index
UNCCD	UN Convention to Combat Desertification
UNDP	United Nations Development Program
UNESCO	United Nations Educational, Scientific and Cultural Organization
UNSO	Office to Combat Desertification and Drought of the United Nations Development Programme

Introduction

Linda Courtenay Botterill and Melanie Fisher

Drought is a normal feature of the Australian farmer's operating environment. This apparently unremarkable statement underpins much of what follows in this volume, although it was not until 1989 that this principle was introduced into the national drought response—and its implementation remains problematic. As has been eloquently described elsewhere[1] and is argued by Stafford Smith in the next chapter of this book, Australia is, in the real sense of the word, unique. The forces that shaped this continent were quite different from those that created the fertile soils of Europe, from where we have imported many of our agricultural practices. Importantly in the context of this book, our climate patterns are erratic, unreliable and are not based on an annual cycle—creating problems for agricultural producers, their families and their communities. Of course, drought has impacts on urban Australia and other non-farm communities as well. However, consistent with the approach of Australia's drought policy response to date, the focus of this book is on the effect of drought on agriculture, farming communities and the responses of the rural policy community.

The frontispiece to this volume shows the distribution of El Niño and La Niña events, and significant droughts over the last 100 or so years, illustrating the variability confronting farmers and the extremes of both wet and dry years. In this environment the concept of an 'average' year is little more than a mathematical construct. How do Australians respond to this level of variability? Was our pre-1989 approach of treating drought as a natural disaster appropriate? How do we ensure that the drought policies we have in place are not setting the scene for further land degradation in an already biophysically stressed continent? How do we respond equitably to the needs of farm families faced with prolonged or extreme drought, while simultaneously ensuring the survival of the productive basis of Australian agriculture and preserving our fragile environment? How do we ensure that the policies we have in place provide support for those in need, but do not sustain otherwise unviable or bad practices with the associated costs to the economy and the environment?

In considering many of these issues, this volume raises more questions than it answers—which is part of its intention. We have sought in compiling these essays to illustrate the complexity of drought and the huge task confronting decision-makers seeking to develop a sound, sustainable policy response. The chapters are arranged as follows. We start with something of a bird's-eye view of our landscape and its climate. We then present a brief history of government responses to drought to explain where we are in policy terms and how we got there. We follow this with a discussion of the reality of drought, the types of policy instruments available to governments and the implications of these instruments. We conclude with an international perspective on drought policy. The book is flavoured with the different disciplinary perspectives of our authors and we hope that this contributes to an understanding of the need to think laterally when considering drought and its impact.

Learning to be Australian

One of the key themes of this book is the need for agricultural producers, governments and the broader community to understand more fundamentally what it means to 'be Australian'—working with the biophysical resources and climate patterns of this continent, rather than imposing ideals and practices more suitable elsewhere. Learning to be Australian also means living with the socioeconomic characteristics of this country—its three tiers of government, its national identity, which continues to be closely tied to the bush in spite of our highly urbanised population, and the limited budgetary resources available to a population of barely 20 million.

While there is a growing awareness of the challenges of working with the Australian environment and of the degradation that results from inappropriate land management practices, it is our view that there remains a need for increased awareness among policy-makers, farmers, the media and the broader community of what it means to operate in an Australian environment. Issues such as water allocation and appropriate pricing of this scarce resource should be the basis of important national debates. Many Australians are not aware that the cheap food we enjoy is a result partly of the environmental externalities associated with agriculture for which no-one pays at the moment, but which will impose costs on future generations. Learning to be Australian means understanding the foundations on which we are building our agricultural systems in biophysical terms and developing a sophisticated climate literacy that brings to an end a tradition of stunned amazement at the onset of drought. Ideally, the term 'drought' itself should be struck from the national language and replaced with 'climate variability'—or perhaps not be the subject of discussion at all!

Drought is complex and multi-faceted

At first blush, the concept of drought seems to be quite straight-forward—it is the result of a lack of rain. However, there is no agreed definition of drought. It can be meteorological, hydrological, agricultural and/or socioeconomic.[2] Fundamentally, drought occurs when there is a mismatch between the water available and the demands of human activities.[3] As human communities expand and develop, they tend to move into more and more marginal areas, a phenomenon Glantz has described as moving 'down the rainfall gradient' on to land that is less suited to agricultural production,[4] giving the impression that drought is increasing in its frequency as its impact increases.

In Australia drought cuts into our national income, puts enormous stress on our farming families and communities and, when combined with some farming practices, can degrade our environment. Designing a policy that can respond to all of these pressures is a major challenge as the response to one of these elements can exacerbate another. Any policy will inevitably involve trade-offs between conflicting objectives.

There is no single correct drought policy

In a democracy, arriving at the balance between these objectives is the role of our elected policy-makers. They have the unenviable task of striking the balance between the different needs of the social, environmental and economic systems. This book aims to highlight some of the important issues our authors believe policy-makers need to take into consideration when making their decisions. We hope to stimulate the reader to ponder what policy processes will deliver drought policy that is in the best interest of the broader Australian public, rural communities and farmers, in terms of our collective and individual values, and that will sustain the resource base. We do not have the answer, but it is our firm view that an informed public debate starts with an informed public. This volume attempts in its modest way to begin that debate.

After this introduction, Mark Stafford Smith sets the scene for our discussion of drought in Australia. Chapter 1 provides a thoughtful overview of Australia and its biophysical characteristics. It outlines the socioecological factors that the author considers need to be accounted for in the development of drought policy, including recognition of the political and social constraints within which policy-makers operate.

This leads on to an analysis by Janette Lindesay in Chapter 2 of Australia's climate, the influences that shape our climate, and the nature of drought and its impacts on Australian agriculture. This understanding of the science of climate plays a critical role in attempts to adapt the country to suit our needs—and in adapting human activity to the realities of the Australian climate. Two of the key messages emerging from this chapter are that droughts are a normal part of the climate experience over much of the continent, and that although there is still some way to go, considerable advances have been made in understanding climate and the implications of these realities for agriculture.

Chapter 3 by Linda Botterill outlines how governments have responded to drought in Australia, with a particular focus on policy developments since 1989. The author considers the policy-making process itself and how this can impact on the types of policy instruments employed. The constraints imposed by the federal system and by democracy itself as drought becomes a political issue and politicians are faced with increasing pressure to respond are examined. Attempts by policy-makers to ensure consistency between drought policy and governments' structural adjustment objectives for the rural sector are discussed. Through a discussion of the conflicting priorities that confront policy-makers and the different outcomes that result from a reordering of those priorities, the chapter picks up the argument that there is no single correct drought policy waiting to be discovered.

An important ingredient in the mix that drives politicians to act in response to drought is the media. On the basis of years of experience in reporting rural issues, Åsa Wahlquist examines the media's reporting of drought and its impact on public perceptions in Chapter 4. She takes as her case study the drought of the early 1990s, which was setting in as the new policy paradigm of drought as risk was being introduced. In this chapter the way in which the media reports drought and the impact of that reporting on the broader community and politicians is explored. Wahlquist's findings are ambivalent. She concludes that the media can have a very powerful role in influencing public opinion and, by extension, policy direction, but suggests that such power is not used wisely— focusing as it does on a disaster approach to drought, and avoiding the more difficult and challenging issues relating to sound farm management and environmental sustainability.

The social consequences of drought can be devastating for farm families and their supporting communities. Although the human impacts of drought are frequently the subject of heated public debate and significant media attention, there is surprisingly little research on the Australian experience of living through drought. In Chapter 5, Daniela Stehlik provides a thoughtful and thought-provoking analysis of how drought impacts on people and families; what influences the capacity of individuals, families and communities to cope with drought; and what lessons policy-makers might take from a better understanding of the social aspects of drought. Stehlik's analysis is informed by data collected from farm families in Queensland and NSW during the 1996–1998 drought period. This research provides valuable insights into issues around how people respond to and recover from drought—and it also provides a voice to the people living the experience of drought.

In Chapter 6, Bruce O'Meagher provides an introduction to the economic perspective on drought in Australian agriculture, the economic arguments around the role of governments in responding to or managing for drought, and drawing on the experience of the 1990s drought, identifies some policy lessons.

Mark Stafford Smith, in Chapter 7, examines the linkages between the biophysical elements that create drought, the social factors that structure producers' experiences and responses to drought, and the policy environment that helps define those experiences and responses. Drawing on coupled biophysical and pastoral decision-making models and interviews with pastoralists, he seeks to describe the implications of evolving human-environment systems. The author concludes with a discussion of the implications for future drought policy.

In Chapter 8, Peter Hayman and Peter Cox explore the issue of risk management and drought in more depth. The post-1989 approach to drought in Australia has been based on the concept that drought is a business risk to be managed by the farmer like any other—such as interest rate, price or exchange rate fluctuations. But do farmers who are living through a drought see it as a risk that can be managed? Drawing on literature from agricultural science, economics, sociology and psychology, the authors address the risks associated with periodic drought, how farmers perceive them, how scientists try to quantify them and how quantification can be used to improve risk management. Given the social and psychological nature of risk, Hayman and Cox urge caution about relying too heavily on apparently 'objective' measures of risk management as the answer to drought preparedness and response.

Don Wilhite, in Chapter 9, places Australia's experience with drought preparedness and management in an international context. He describes international trends in drought policy and how these have been implemented in the United States (US), sub-Saharan Africa and Australia. Wilhite emphasises the importance of regional approaches and the need for international cooperation and information-sharing—including through a Global Drought Preparedness Network. In this chapter Wilhite illustrates that Australia's drought policy, along with that of the US, is somewhat ahead of the pack in implementing agreed principles of risk management and drought preparedness.

We conclude in Chapter 11 by revisiting the themes of the book.

---

1 Flannery 1994.

2 Wilhite and Glantz 1985, p. 113.

3 Dracup *et al.* 1980, p. 297; Wilhite, D. A. 2000b, p. 16.

4 Glantz, 2000, p. 290.

1

Living in the Australian environment

Mark Stafford Smith

For 100 years after the Europeans arrived in Australia, their painters portrayed a landscape that harked back to the soft lights of Europe, their poets wrote wistfully of their respective mother countries, and their music was folk songs from rural Britain. In the context of this culture, rooted in Europe, men of the land (it was mainly men) carried out farming as if it was merely a matter of applying a fine work ethic to subdue the country into a reliable European image.

Around the end of the nineteenth century, artists began to portray a harsher countryside, poets to extol a more local larrikin approach to life, and composers to incorporate images of the bush peculiar to Australia like the kookaburra into their music. In 1894, the first major scientific expedition to central Australia, the Horn Expedition, reflected these changing views.[1] Among its participants there were still those seeing a landscape with European eyes. Others, most notably Baldwin Spencer, had begun to see how the country varied spatially as he and his fellow scientists moved through it on their camels, and over time as he visited repeatedly during the following years. Spencer met pastoralists who were beginning to understand and take advantage of this new land. But his descriptions still make it clear that most inhabitants regarded the ups and downs of climate as being an unfair imposition from on high, rather than a normal feature of the environment, to be managed and celebrated.

In this chapter, the special features of the Australian physical and social environment that affect the way we manage our non-urban areas are outlined. The outback, which has had a deeply symbolic place in the way Australians view their country, today mainly refers to the arid and semiarid interior. What has been viewed as outback has changed over the years—at one time everywhere across the Blue Mountains was included, but the boundary between outback and inside country slowly flowed towards the interior as settlement proceeded. However, issues of drought management affect all non-urban areas and many of the issues that are writ large in the arid zone still affect other regions in a slightly more subdued fashion.

The biophysical environment

Key features of the Australian biophysical environment have been summarised many times.[2] It is easy to generalise across the whole continent, and such statements must always be tempered with the real diversity of more local conditions. This will become important when we come to consider the scale at which issues such as drought should be managed. For example, we inevitably focus on the nature of climate in Australia, but as Figure 1.1 shows, and we all know, there is an immense diversity of climatic patterns represented in the continent, from the relatively reliable monsoonal systems in the north, through to the incredibly uncertain arid centre, to the somewhat more reliable temperate southern systems. We will return to this issue.

At the base of it all, this is an ancient continent, worn into low relief by millions of years of exposure to the changing atmosphere, and concentrated into salt lenses and silcretes by an equal period of leaching. Coupled with the resulting low productivity, at least for the last few tens of thousands of years, the continent has been located in a particularly variable part of the earth's climate system. On top of the normal annual variability found in all semi-arid areas in subtropical to temperate zones of the planet, Australia experiences additional multi-annual variability drivers such as El Niño.[3]

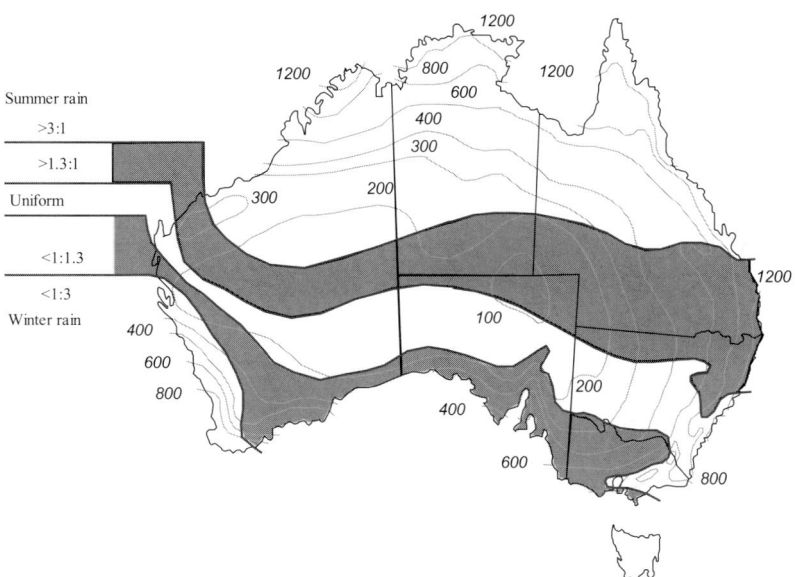

Figure 1.1 Seasonality and median annual total rainfall across Australia (after AUSLIG 1992). Ratio of summer rainfall to winter rainfall.

We are also increasingly recognising that there is an inter-decadal time scale of variability that may be specifically modulated by the 'Inter-decadal Pacific Oscillation', but this is affected by other long cycle events in the world's oceans; and, of course, we may now be facing directional climate trends driven by global climate change. Significant rainfall events are immensely more diverse for an Australian site than for a US site in a comparable mean annual and seasonal rainfall regime.

These features alone have major implications for the Australian biota.[4] Plants living in the US (or Mediterranean) environments can be reasonably confident that they will receive another rainfall of a given size with a consistent return time, usually less than a year. By comparison, the Australian plant (speaking teleologically!) has no idea when its next drink is going to fall. In consequence, stem succulence is a highly successful strategy in plants in Central America, while species with these characteristics are almost entirely absent from Australia.[5] This is not a surprising response to the fact that, while you can store a water supply in your stem with confidence of replacement within a year in America, this would be

11

(or has been) a suicidal recipe in Australia with its uncertain climate. As another example, in a regular climate where you know there will be a time of year through which it is hard to persist, it makes sense to be deciduous—to drop your leaves and reduce your costs of obtaining water at times when it is in short supply. But if you live in Australia and you don't know when these times are going to occur, or when they are going to end, your leaves become a more precious resource. This effect is exaggerated still further if you live on poor soils where nutrients are at a premium, and you don't want to be regrowing your leaves from scratch on a regular basis. For these reasons and others, Australian plants tend not to be seasonally deciduous, though some exhibit drought-deciduousness, having a strategy for slowly dropping leaves as conditions become drier.

Such ecological linkages may seem quite esoteric, but they have major implications for native or domestic animals that live on these resources, and for sustainable land management. Our native animals have come to their own accommodation with these underlying ecological drivers. The classic example often cited is the kangaroo, with its suppressed embryo developments, such that if one joey is aborted there is a new foetus instantly ready to take its place when conditions become favourable. But there is a multitude of comparable relationships that are more subtle, but systemically far more significant; for example, driving the ecosystem services provided to agriculture by ants and termites. Again, these have been discussed in a variety of places, although the way in which their implications flow through to agricultural management is still a matter for considerable debate.

Two immediate examples related to management are fire and grazing. Because plants on poor soils are relatively more constrained by nutrients and by the production of carbohydrate, many Australian ecosystems produce plentiful and long-lived fuel, and are consequently subject to fire. There are many other reasons why fire is an important force in the Australian ecosystems, and why many Australian species may have evolved to 'use' fire, but the result is that management must cope with and take advantage of the effects of fire. Our slow development of understanding about the intimate relationships between fire and vegetation composition in our landscapes has led to a variety of problems. This includes woody weed invasion in rangelands and the loss of resilience in the management of our forest ecosystems, which contribute to or contrast with the issues related to drought management.

Not surprisingly, grazing also interacts with the features of Australian ecosystems. There are a variety of different pasture types within grazed Australia; a significant proportion of these are dominated by native plants that can cope quite well with intermittent catastrophic losses of foliage, but very poorly with chronic defoliation such as that caused by continuous grazing. Thus we have seen widespread loss of palatable perennial plants in Australia's grazed lands. At the same time, while early settlers soon found and exploited those patches of country which had richer soils and which carried plants that were less constrained by nutrients than on the general landscape, we are still learning how to manage this complex mosaic of environments. The problem is that the overall general low productivity of many of our farmed systems means that individual management units (such as paddocks) must be large, and consequently end up including a diversity of landscape types. This makes them much harder to manage than small fields, which can segregate chunks of the landscape with different characteristics and then manage them differently.[6]

Agriculture and forestry too must deal with these underlying characteristics of our ecosystems. Forests contain a mosaic of types of trees in much the same way that rangeland environments contain mosaics of different perennial grasses. Richer patches of soil support trees and grasses that tend to be more productive, and are associated with a particular suite of native animals that require food of higher quality. Just as cows may compete for food by selectively grazing the better quality forage, so selective logging for faster growing types of tree may often compete with a certain suite of native animals for this resource. In both cases, those ecosystems that depend on the richer soils tend to be selectively used, or even over-used, and their dependent ecosystems thereby put at risk. These richer pockets in a generally poorer landscape are also the areas that are most completely cleared for intensive agriculture, so that the same issue arises in a different form in agricultural landscapes.

Why do all these issues matter for drought management? The generality behind them is that the intensification of ecosystem use, whether from native animals to pastoral grazing, or from grazing to intensive agriculture, is a process of gradually replacing the ecosystems' internal buffering dynamics by external management and subsidy. For example, in moving from kangaroos to cattle, we move from a system in which kangaroo populations boom in good years and move or die back in poor years, to a system in which we constrain grazing animals from moving so freely across the landscape, but subsidise their survival in poor years by providing artificial water and even supplementary food. Similarly, in moving to intensive agriculture, we aim to obtain much higher production per square metre than previously, but do so by providing irrigation, fertilisers, and a great deal of mechanical intervention through planting and harvesting. We replace natural but relatively inefficient and unreliable means of moving water and seeds across the landscape with artificial channels and mechanised planting. These developments are a necessary part of obtaining greater production per unit area of landscape, the principal societal goal for some regions, and this must be balanced with appropriate actions to meet other landscape goals such as the conservation of diversity. However, these actions also reduce the intrinsic resilience of the agroecosystem to deal with external shocks such as drought, a fact that we must therefore recognise and explicitly manage for.

Understanding the specific ways in which our ecosystems originally coped with the special features of the Australian biophysical environment is essential in creating the appropriate management regime under which their natural resilience is replaced by management actions. Of course, the creation of such a management regime depends not only on knowledge of the biophysical system, but also on the social and political environment within which the knowledge is to be implemented, to which we now turn.

The social and political environment

Just as the biophysical environment of Australia has features that, while not being individually totally unique, in sum distinguish it from most other regions of the world where our lessons about drought management might be drawn, so too does its sociopolitical environment.

Several centuries ago, Aboriginal inhabitants walked this continent relatively lightly, capitalising on years of plenty while coping with deprivation in years of drought. It was a classic low-input–low-output system, which had probably come into balance with variability in the biophysical environment over millennia of experimentation. People used large areas relatively conservatively. Various cultural rules were developed for protecting those resources that were critical in dry times, through the times of plenty when people might have been tempted to overuse them. While the demands of much higher population levels and higher expectations for standards of living mean that European-style, intensified land use is here to stay, there are lessons that can be learned from the ways in which the indigenous population created resilience in its interactions with the Australian landscape. To make such a case we must first look at the features of the Australian sociopolitical landscape that continue to distinguish it from other regions in the world.

First, and most obviously, the non-urban areas (whether just west of the dividing range or, more extremely, out in the arid zone) of Australia are relatively sparsely populated.[7] This reality has a whole series of downstream implications. Markets tend to be more remote or smaller, resulting in higher production and transport costs. Historically, remoteness has bred both a degree of self-reliance, but also a certain degree of disdain for the other end of the market chain, resulting in relatively poor feedback in terms of clients' concerns. The sparse population also means that it is relatively difficult for local groups to get together, to act together, or to exchange information and understanding about how their systems work. In the past decade the Landcare movement has begun to evolve a missing level of local community governance in Australia, but with mixed success, particularly as regions become less densely populated. The limited communications technologies of the past have also tended to cause a disjunction between rural communities and the views of their urban peers, particularly in terms of growing concerns about environmental issues. Recent dramatic changes in communications and access to media are rapidly changing the situation, but it has held true for the vast majority of Australia's European history.

Second, the Australian bush ethos has evolved with a curiously schizophrenic history. On the one hand, bush people have been fiercely independent and sceptical of the motives of central government. On the other hand, governments have long promulgated and, with varying levels of effectiveness, implemented the intent of settling the 'empty continent'. The almost mythological place that the outback has in the heart of urban Australia, enables a small rural electorate to have a disproportionate influence on the political process through the emotional ties of the urban populace. As we march into a period in which some regions are clearly moving in to a post-productivist future based on non-market values, while others stay with the productivist paradigm of the past[8], this schizophrenia is not declining.

Third, our federal political system creates additional oddities that any natural resource management institutions system must deal with. The responsibility for natural resources is vested in the individual states, which consequently replicate an immense amount of bureaucratic structure in order to implement laws for fundamentally similar purposes in each State. The Commonwealth, on the other hand, can only intervene through a limited number of instruments, including taxation and the provision of funding with cross-compliance requirements on some natural resource management issues. This is peculiar because, inasmuch as natural resource management issues need to be dealt with in a way that is sensitive to local conditions, this is needed at a regional scale. Ideally, broad-brush policy is consistent across the nation; however, this needs to be negotiated on a federal scale. The state scale is intermediate and not particularly well-suited to meeting either purpose. The result is inconsistencies and conflicts, trivial or otherwise, across state boundaries, and enormous transaction costs for obtaining cross-border solutions. One would think that for many purposes a more satisfactory solution would be obtained through national policy-making with regional implementation.

To return to the example of the continent's earlier inhabitants, while we clearly cannot go back to the population densities and lifestyle hardships (by current standards) of those days, it is worth noting that their resilience to climatic variability in this land was sustained by an ability to obtain their needs from a large spatial area, by treating their core resources very conservatively, by the possession of intimate local knowledge, and through high social capital in the community. Each approach has its place in today's natural resource management. Policy can facilitate commercial or other arrangements that enable farmers to spatially hedge their exposure to drought, through cooperative arrangements with other regions; the significance of the many such arrangements that farmers already have for this are often overlooked. A balanced exploitation strategy in which core resources are not damaged remains crucial, and supporting social capital and local knowledge at the appropriate scale rather than imposing supposed national solutions remains highly relevant.

In the remainder of this volume, we will explore issues of this type in greater detail. I complete this chapter by outlining some immediate general ways in which an understanding of the socio-ecological environments of Australia should inform future drought policy solutions.

Implications for drought policy

Drought policy must deal with at least two different types of issues. First, policy aims to facilitate the appropriate type of on-ground management; this implies that policy-makers must understand (in general terms, at least) what sort of management is required to respond to the biophysical conditions that managers are operating within. Second and equally important, policy instruments are themselves an institutional response to a problem, and policy-makers need to understand how to design the most effective types of institutions to obtain the intended on-ground outcomes.

Today, there is a growing body of theory and practice around the concept of learning communities and institutions.[9] Institutions (or more correctly, the group of people operating within them) can only learn if they, first, are able to detect and attribute the positive and negative changes that their actions are causing and, second, have the knowledge, motivation and capacity to adjust their actions and structures in the appropriate way. This idea can be applied both at the scale of understanding on-ground management, and to the policy-making process itself. The goal is to have both land management practices and policy environments that are resilient, in the sense of being able to cope and evolve with changes and shocks from each of their respective external environments. In meeting such a goal, the specific local environmental conditions are crucial in understanding the local natural resource management responses;

and the continental-scale diversity of environments together with their sociopolitical environments are crucial in informing the policy response. In the remainder of this volume, we will address these issues in a variety of ways, but the following points contain some examples of the general implications of the biophysical and sociopolitical environments that are particular to Australia.

Local–regional scale adaptive system issues

At the local scale, there are a number of key environmental issues:

- Australian agro-ecosystems generally have to cope with high levels of climatic variability at all scales from inter-annual to inter-decadal and longer; some regions, particularly across the centre of the continent, have an added element of uncertainty at the intra-annual level; that is, not knowing when in the year rain is likely to come. Policy intervention needs to be aware of the many different local ways in which climatic variability plays out and affects managers across the continent.

- Many environmental problems that have emerged from our management of Australian agriculture have very long lead times; they are hard to detect in their early stages and hard to attribute.[10] Inasmuch as policy seeks to create an institutional environment in which managers learn effectively how to cope with climatic variability, it must be recognised that this feedback loop is difficult at any time and misguided interventions can very easily reduce such tenuous signals. Policy must also facilitate the ability of managers to monitor the biophysical outcomes of their own actions, since these are hard to detect (and hence to learn from) in a variable climate.

- Many, but not all, Australian ecosystems are relatively unproductive compared to the expectations of the farmers using them; as a consequence, if the system is damaged there is limited capacity to invest in recovery. Understanding this aspect of system resilience is important, since it means that avoiding damage in the first place is more important in some systems than others.[11]

- A core concept in resilience theory is that it is underlying 'slow' variables which are important issues in determining system resilience, not the superficial 'fast' variables. The 'fast variables' that humans depend on in their day-to-day experiences are very real issues for short-term humanitarian aid, but confuse the strategic debate about sustainable natural resource management. Droughts bankrupt families when those families live on eroded landscapes with no stored capital, whether social or economic; the same 'drought' may hardly be noticed by a group of farmers with healthy pastures and low debt. In seeking to support long-term change towards sustainable livelihoods, it is essential to focus on the 'slow variables' that set this context.[12] In this sense, drought simply brings some underlying critical structural problems to a head.

Any policy solutions must address these local issues, but through a framework that can take account of some broader structural concerns.

State–national scale adaptive system issues

At the policy formulation level there is another suite of more institutional issues:

- The policy system needs to be able to learn and adapt to changes in the rest of the socioecological system, just as much as individual farmers must. This may mean that it is important to design a system that can be sensitive to local, regional conditions. This is much easier in a Landcare-type self-reliance institutional arrangement than through national taxation or reconstruction instruments. However, the national context for this system also needs to be designed in such a way that it can evolve constructively over time.

- For the system to evolve over time, there must be feedback data on the actual outcomes (rather than inputs or even outputs) of policy actions, and a mechanism for responding to this feedback without loss of institutional coherence. It is therefore very important that the detailed instruments of any policy, which may need to change over time, are embedded within a well-articulated, broader policy philosophy that does not bounce around over time.

Climate variability is by definition a probabilistic beast. Policy success should not be judged on individual events. It requires a long-term view, which does not sit easily with the electoral cycles if the policy interventions are too hands-on.

In this chapter I have outlined environmental factors that need to shape the philosophy and implementation of any drought policy in Australia. I have raised more questions than answers, but these may be sought in later chapters.

---

1 Spencer 1896.

2 For example, Beadle 1981; CSIRO Division of Soils 1983; Friedel *et al.* 1990; AUSLIG 1992; Groves 1994; Stafford Smith 1994; NLWRA 2002.

3 Nicholls and Wong 1990.

4 See, for example, Barker and Greenslade 1982; Dodson and Westoby 1985; Saunders *et al.* 1990; Stafford Smith and Morton 1990.

5 Stafford Smith and Morton 1990.

6 Stafford Smith 1994.

7 For example, Haberkorn *et al.* 1999.

8 Holmes 2002.

9 For example, Robbins *et al.* 2002.

10 For example, Stafford Smith *et al.* 2000.

11 Stafford Smith *et al.* 2000.

12 Stafford Smith and Reynolds 2002.

2

Climate and drought in Australia

Janette A Lindesay

Australia has an inherently variable climate, characterised by marked fluctuations, particularly of rainfall, in both space and time. The high degree of both seasonality and interannual variability is due largely to Australia's geographic location and the associated atmospheric and oceanic influences to which the continent is exposed. One consequence of a variable climate with distinct seasonality in rainfall and relatively high dry-season temperatures is an environment that is prone to periods of below-average rainfall and, in extreme conditions, to the prolonged rainfall deficits that produce drought.

The environmental history of Australia has been discernibly shaped by the climatic extremes of rainfall and temperature. The degree of adaptation of the biosphere to extremes of climate, and particularly to dry or drought conditions, is an indication that the general patterns of weather systems, climate and the seasons have existed in forms recognisably related to modern patterns since the continents reached their present geographic positions millions of years ago.[1] Human management with, and adaptation to the climate and environment of, Australia is a more recent development, with the arrival of Aboriginal peoples dating to approximately 40 000 to 60 000 years ago and that of European settlers to the late eighteenth century.[2] The history of European responses to Australia's climate, in particular, and the environmental impacts of those responses, have contributed to current landscape conditions and to many of the attitudes and values that we currently bring to managing with a variable climate. In the words of Blench and Marriage:

> *Climate is often conceptualised as a series of shock events punctuating a background of acceptable variation. Shocks, such as floods, high winds and drought, are discontinuities that are sufficiently anomalous within the lifetime of observers as to be classified as unpredictable and life-threatening. The nature of the discontinuity is framed by the region's ability to cope. ... Vulnerability to weather is a function of preparedness as well as of the event in itself.*[3]

The purpose in this chapter is to establish the climatological context for a discussion of drought in Australia, beginning with the characteristics of the climate of the continent, and considering the nature and possible causes of climate variability in the region. The problem of defining drought is addressed, and drought definitions are considered across a range of sectors. The importance of using our understanding of past climatic variability, and drought in particular, to inform both current decision-making and planning for the future is highlighted in a discussion of scenarios for possible future climate change in Australia, and what those changes could mean for the future of drought in the region.

Characteristics of Australian climate

The climate of the Australian continent is determined by the geography of the land mass, which extends from the tropics in the north, through the subtropics to the fringes of the midlatitudes in the south. Each of these broad latitude zones is characterised by different dominant weather systems. In the tropics (from the equator to approximately 20 degrees South), easterly airflow and tropical disturbances (including tropical cyclones), and in some areas the wind reversals and distinct dry and wet seasons associated with the monsoon, dominate the rainfall regime. The subtropics (between about 20 and 40 degrees South) are dominated by anticyclones with slowly descending air masses and clear skies, disrupted by heat lows and tropical disturbances from the north and east in summer and by midlatitude cold fronts from the south and west in winter. In the midlatitudes (poleward of 40 degrees South), a succession of frontal low pressure systems moves across the fringes of the continent in the prevailing westerly airflow, each front bringing a sequence of weather conditions, including warmer, dry air masses ahead of the front and cooler, moist air with the potential for rainfall and possible thunderstorms with the passage of the front.[4] These climatic zones play a fundamental role in shaping weather, climate and climatic variability over the region.

Topography is also an important factor in determining regional climatic conditions in Australia. Although much of the continent has relatively low relief, the mountains of the Australian Alps in south-west New South Wales and north-east Victoria, and the Great Dividing Range (or escarpment) inland of the east coast have at least a regional influence on climate. This is reflected in annual average rainfall that is higher on the windward (coastal) sides of the higher altitude areas than it is on the inland sides; moist onshore airflow that is forced to rise above these mountains frequently produces cloud formation and precipitation, leaving moisture-depleted air to continue inland.

The large west–east extent of the continent means that inland Australia has generally drier conditions and more extreme temperatures than the coastal margins; the oceans generally have a moderating effect on temperature (due to their large capacity to absorb heat), and are often a source of moisture through evaporation. Areas in the interior of the continent, which lies largely in the subtropics, are remote from the sources of moisture-bearing winds, particularly tropical easterlies, crossing the coast. The interior also experiences greater extremes of temperature both diurnally (that is, the difference between daily maximum and minimum temperatures is relatively large) and seasonally (that is, between summer and winter) than do the coastal fringes. The diurnal temperature range is larger in the subtropics, because the generally clear skies and dry air allow maximum heating of the surface and overlying atmosphere during the day, and maximum loss of heat by long-wave radiation from the surface at night.

The result of Australia's geographical location and topography is a pattern of annual average rainfall and temperature with a general gradient from warm, monsoonal tropics to cool midlatitudes, and wetter conditions around the northern, eastern and southern coastal fringes. The interior is relatively hot and dry, with potential evapotranspiration greatly exceeding surface moisture availability; in central Australia near Alice Springs, for example, annual average rainfall is between 200 and 300 mm, while annual average potential evapotranspiration exceeds 1300 mm. These characteristics, and their expression in the environment (in vegetation types, for example, which integrate the combined effects of moisture availability and temperature) have been summarised in classifications of climate, of which the best known is probably the Köppen Climate Classification (see Figure 2.1). An important aspect of this and many other climate classifications is seasonality, since the timing of rainfall relative to the annual temperature cycle is important for the biosphere.

Seasonal variations in weather and climate occur principally as a result of the migration of the overhead sun between the tropics of Cancer and Capricorn, reaching a northern limit at the winter solstice in June and a southern limit at the summer solstice in December. The consequent winter expansion of cold Antarctic air masses and northward movement of midlatitude low pressure systems produces winter rainfall over southern Australia (for example, Perth, Figure 2.3a); and the expansion of warm, humid tropical air masses southward in summer brings tropical disturbances and moisture sources further south over the continent (for example, Sydney, Figure 2.3b).[5] Across northern Australia the seasonal migration of the zone of maximum convergence of airflow and cloud formation between the northern and southern hemisphere tropics leads to the wind reversal (moist south-easterly in summer, dry north-westerly in winter) associated with the monsoon (for example, Darwin, Figure 2.3c). These atmospheric circulation characteristics mean that there is some degree of seasonality in rainfall almost everywhere in Australia. Differences between January (summer) and July (winter) rainfall are clearly greatest in the monsoonal tropics and in the south-west winter rainfall region, and seasonal temperature differences are large in the subtropical interior, the south-west and the south-east. The influence of topography on both temperature and rainfall is particularly evident in south-eastern Australia.

The seasonal characteristics of Australian climate are average conditions, however. As is typical of the global subtropics, much of the continent is subject to a marked degree of interannual climatic variability (see Figure 2.2). The weather and climates of both the tropics and the midlatitudes may be affected by large-scale circulation fluctuations from year to year that can affect the location, frequency and intensity of particular types of weather systems. Any changes in these two zones will impact on the subtropics, which are thus particularly prone to interannual variability as expressed in extremes of rainfall and temperature. The impacts of year-to-year fluctuations in rainfall, in particular, can be large in regions characterised by distinct dry and wet seasons; a failure of rain during the normally wet season means that no effective falls can be expected until the start of the next wet season. This can result in prolonged periods of below-average rainfall and drought.

Changing rainfall seasonality

Recent research on changing seasonality in Australia's rainfall[6] has highlighted the spatially varying nature of long-term fluctuations in monthly rainfall across the continent. Probably the most important aspect of these fluctuations is the changes that have occurred in rainfall in preferred months of the year, while rainfall in other months has changed little during the more than 100 years of meteorological record. At Sydney in south-eastern Australia, for example, since 1860 there has been an increase in rainfall in January and a decrease in July, both of the order of 30 per cent of median monthly rainfall for those months. The result is that at Sydney, and at many other places in Australia, rainfall seasonality is not fixed, but varies between periods of well-defined, high-amplitude seasons, and periods when rainfall seasons are poorly defined and the transitions between seasons may be blurred. The fact that the seasons are not fixed in either amplitude or timing, and that the start and end of the wetter season in subtropical Australia has varied by at least a month during the last 100 years or so, adds to the complexity of identifying drought. These fluctuations in rainfall are not random, however, and are apparently related to some of the most significant large-scale climatic causes of low-frequency rainfall variability in Australia.

Causes of climate variability

Perhaps the most widely recognised large-scale influence on inter-annual rainfall variability in the global subtropics is the El Niño Southern Oscillation (ENSO),[7] which is particularly important in modulating rainfall variations from year to year across much of eastern Australia. The ENSO phenomenon is the largest known interannual fluctuation in the ocean-atmosphere system. It centres on the tropical Pacific Ocean, although its characteristic patterns of atmospheric pressure, winds and temperature extend into the Indian Ocean region, and its impacts are near-global. ENSO involves a suite of interlinked anomalies in both atmosphere and ocean, most obviously seen in sea surface temperature variations across the tropical Pacific Ocean.

The lowest air pressures tend to be co-located with the highest sea surface temperatures in the tropics, so that on average there is a low pressure centre (the Indonesian Low) over the 'maritime continent' area north of Australia, where sea surface temperatures are high. The South Pacific Anticyclone (a high pressure centre) is associated with the cold sea surface temperatures along the South American coast in the eastern South Pacific. The south-east trade winds blow across the tropical Pacific between these two pressure centres, from the northern fringe of the anticyclone towards the low pressure area. The convergence of moist airflow near the surface into the low pressure area creates uplift (convection) and cloud formation in the tropics, carrying energy upwards in the atmosphere. This energy is then transferred eastward (across the Pacific), westward (over the Indian Ocean) and southward at altitudes of about 10 km, before the air sinks towards the surface in the semi-permanent anticyclones of the subtropics. These large-scale overturning features of the general circulation of the atmosphere are known respectively as the Walker (east-west or zonal) and Hadley (north-south or meridional) circulations. Any disruption to the 'normal' state of the atmosphere and ocean across the tropical Pacific Ocean can thus lead to changes in winds and rainfall-producing weather systems throughout the tropics and into the subtropics and midlatitudes, via changes to the large-scale atmospheric circulation.

Changes in the gradient in sea-level pressure between the two principal Pacific basin pressure centres are known as the Southern Oscillation, and are used as a measure of the status of ENSO with a variety of indices. The best known of these is the Tahiti–Darwin Southern Oscillation Index (SOI).[8] The SOI is calculated as the difference in monthly average sea-level pressure between Tahiti (in the vicinity of the South Pacific Anticyclone) and Darwin (in the Indonesian low pressure area). Each month's pressure value is standardised to remove the seasonal cycle, which would otherwise dominate the record. The SOI calculation used in Australia follows the method of Troup, which provides typical SOI values between +20 (La Niña events) and –20 (El Niño events); the pressure gradient measured by the SOI is a continuum, however, and is more often closer to 0 (normal) than in either extreme state. Time-series of the Tahiti–Darwin SOI can be calculated from 1876[9] and have been correlated with rainfall in many parts of the world, including Australia. Other, more accurate, indices of ENSO activity have been developed using shorter data series; among these are the Multivariate ENSO Index (MEI) and the Niño 3.4 sea surface temperature Index (based on sea surface temperatures in an area of the central equatorial Pacific), both of which can be used to track ENSO since the 1950s.

During the El Niño extreme of ENSO, above-average sea surface temperatures extend from the South American coast into the central equatorial Pacific, and temperatures over a large area may rise to more than 6°C above normal. At the same time, sea surface temperatures around the north-east Australian coast are often cooler than normal. This represents a significant change to the normal east-west temperature gradient across the tropical Pacific (which is generally warmest in the west, that is north of Australia, and coldest in the east, that is off the South American coast). Sea-level pressures rise across the western equatorial Pacific and northern Australia, and fall in the area of the south Pacific Anticyclone. This leads to a decrease in the gradient of pressure and thus a decline in strength of the tropical south-easterly trade wind flow. The principal atmospheric low pressure area and convergence zone in the tropics, normally located over the warm waters north of Australia in summer, moves eastward to the vicinity of the international dateline, where the area of above-average sea surface temperatures has developed. Tropical convection and moist easterly onshore airflow diminish over Australia, and parts of eastern Australia are thus generally sunnier, drier and warmer than normal during El Niño events (see Figure 2.4a). These events recur on average every 3–4 years,[10] although the return period varies; for example, there were four successive El Niño years in the period 1991–1995, followed by another event in 1997–98. The most recent El Niño event occurred in 2002–03, and was associated with one of the most severe drought periods on record in parts of eastern Australia.

Wetter than average conditions typically occur in eastern Australia during La Niña events (see Figure 2.4b), when tropical Pacific sea surface temperatures are colder than usual. There is a tendency in the climatological record for La Niña conditions to follow El Niño events, although this is not always the case. This contrast could account, at least in part, for Australia's reputation as a land of droughts and flooding rains.

Despite its well-described behaviour and general impacts, ENSO is highly variable; no two El Niño events have the same evolution, intensity or impacts. This makes precise prediction of ENSO impacts particularly difficult, as is exemplified by the moderate (although spatially coherent and physically meaningful) overall correlations between ENSO and rainfall in Australia. It also means that indices of ENSO activity such as the SOI must be used cautiously in a predictive context. ENSO is not the only influence on rainfall in Australia, and not all droughts can be ascribed to El Niño events. Other contributing factors include the Pacific Decadal Oscillation,[11] the Indian Ocean Dipole or Dipole Mode Index,[12] and the Antarctic Circumpolar Wave.[13]

Recent research has highlighted the potentially important role of the low-frequency fluctuations of the Pacific Decadal Oscillation in modulating ENSO influences on Australian rainfall.[14] The phenomenon, which has characteristic patterns of sea surface temperature across the northern Pacific Ocean, varies slowly on decadal to multi-decadal time scales. When the Pacific Decadal Oscillation is positive, ENSO events are more likely to impact moderately on Australian rainfall; when the Pacific Decadal

Oscillation is negative, there is an increased likelihood of ENSO events impacting strongly in Australia. The Pacific Decadal Oscillation, which was positive between around 1980 and 2000, has since entered a negative phase. Thus the background climatology for the 2002–03 drought season, with an El Niño event occurring in a negative phase of the Pacific Decadal Oscillation, may have assisted in establishing the conditions for severe impacts of a relatively moderate El Niño event (as measured by the SOI and MEI) in Australia.

Understanding the causes of climatic variability, such as the significant role of the oceans and sea surface temperature variability (as exemplified by ENSO and the Pacific Decadal Oscillation), enhances our ability to unravel the complex interactions between the impacts of natural fluctuations in the system and those impacts that may be ascribed to human activity. Physical understanding of the nature of climate fluctuations and their impacts is also contributing to seasonal climate forecasting, which can be a useful tool for land- and water-resource managers and in agriculture. Knowledge of forthcoming drought conditions will not permit those conditions to be avoided, but does allow appropriate management decision-making to minimise drought impacts.

Aridity and drought in Australia

As has been shown, much of Australia lies in the semi-arid subtropics, with average annual rainfall totals below 350 mm per year. Long-term average rainfall is calculated as the arithmetic mean of annual rainfall at each location over a period of time, usually at least 30 years (the most recent World Meteorological Organisation climatological averaging period is 1961–1990). However, in regions of high interannual variability such as the subtropics, the long-term average may not reflect the actual nature of rainfall in any particular year. Identifying drought can be particularly difficult in these areas.

The concept of drought

Drought is a normal and recurrent feature of climate; it has occurred everywhere on Earth, but extended drought is more likely to be observed in regions of high interannual rainfall variability. The term 'drought' implies a lack of precipitation over an extended period of time, such that there is insufficient moisture for a particular region, land use or environmental sector. Moisture deficits are often exacerbated by high temperatures, low humidity and stronger than average winds, which increase evapotranspiration and the impacts of drought. One of the characteristics of drought, as distinct from most other natural hazards, is that drought conditions develop gradually and cumulatively as the balance between precipitation and evapotranspiration worsens progressively due to the failure of 'normal' rainfall. Thus the onset of drought conditions is often difficult to determine. The end of a drought may also be difficult to define, since the effects may linger for some time after precipitation increases again. Managing the impacts of drought is complicated by the fact that they are often widespread both geographically and across a range of economic sectors, and do not necessarily involve structural damage, making them difficult to quantify.[15]

Drought is distinct from aridity, which is a permanent condition of low rainfall. However, the exact definition of drought is elusive; Wilhite and Glantz[16] surveyed more than 150 drought definitions, and there are many more. Any useful definition of drought should be specific to the particular region and application under consideration. For example, regions where rainfall is generally frequent and relatively reliable may develop water deficit conditions after only weeks without rain, whereas in areas with pronounced dry seasons and high interannual rainfall variability a season or more of below-average rainfall is required for a water deficit to develop. In terms of applications, the water requirements of rain-fed agriculture and of hydrological systems are quite different, with the former more rapidly influenced by drier than normal conditions, particularly at critical times in the cropping cycle. And areas of high water usage are more readily vulnerable to water shortage than those where water is used less intensively, highlighting the role that humans play in influencing and defining drought.

All definitions of drought are human constructs: 'drought' or 'flood' are determined as departures from perceived 'normal' conditions; and the definition of what constitutes normality is itself not absolute, since it depends on the range of variability that has occurred during the period of record used to determine that normality. For many people, human memory defines what is considered an extreme event. Scientifically, most climatic records cover no more than the last 100–150 years, so that definitions of extreme events have been derived in the context of century-long climate variability. In recent decades, records have been extended backward in time using palaeoclimatic reconstruction techniques, which has allowed a broadening of the scientific view of what constitutes 'normal' climatic variability to millennia rather than centuries. The perspective that is used in defining drought must depend on what is appropriate to the activity, time and place under consideration, so that exposure and vulnerability to drought impacts can be assessed.

Drought definitions

The *concept of drought* is best understood when definitions are given in general terms. Thus an agricultural definition of drought might be that:

> *Drought is a protracted period of deficient precipitation resulting in extensive damage to crops, resulting in loss of yield.*

The conceptual definitions of drought for applications such as hydro-electric power generation in New Zealand, or wildlife management in southern Africa, or urban water provision in the United Kingdom, would all be different. One common factor in all drought definitions should be the incorporation of an under-standing of climate variability. The conditions under which governments would provide financial assistance to those affected by drought impacts are those beyond what could be considered part of 'normal' climate variability and risk management. Declarations of what might be termed 'exceptional' drought are then based on scientific assessments, bearing in mind that these are also not absolute.

Operational definitions of extreme events assist in identifying the beginning, end and degree of severity of the event. To determine the beginning of drought, operational definitions specify the degree of departure from the average of precipitation or some other climatic variable over some time period. This is usually done by comparing the current situation to the historical average, often based on a 30-year period of record. The threshold identified as the beginning of a drought (for example, 75 per cent of average precip-itation over a specified time period) is usually established somewhat arbitrarily, rather than on the basis of its precise relationship to specific impacts.

An operational definition for agriculture might compare daily precipitation values to evapotranspiration rates to determine the rate of soil moisture depletion, then express these relationships in terms of drought effects on plant behaviour (that is, growth and yield) at various stages of crop development. A definition such as this one could be used in an operational assessment of drought severity and impacts by tracking meteorological variables, soil moisture, and crop conditions during the growing season, continually re-evaluating the potential impact of these conditions on final yield. Operational definitions can also be used to analyse drought frequency, severity and duration for a given historical period. Such definitions, however, require weather data on hourly, daily, monthly, or other time scales, and, possibly, impact data (for example, crop yield), depending on the nature of the definition being applied. Developing a climatology of drought for a region provides a greater understanding of its characteristics and the probability of recurrence at various levels of severity. Information of this type is extremely beneficial in the development of response and mitigation strategies and preparedness plans.

Perspectives on drought

Meteorological drought is defined on the basis of the degree of dryness (in comparison to some average amount of rainfall) and the duration of the dry period. Meteorological drought must be defined regionally because the climatic conditions that result in below-average precipitation vary; for example, drought could be defined in terms of days without rain in the wet tropics, but in terms of months or seasons without rain in the seasonally arid subtropics. Thus periods of meteorological drought are identified on the basis of relating actual precipitation departures from average amounts on monthly, seasonal or annual time scales, as appropriate for the region under consideration.

Hydrological drought is defined in terms of the effects of below-average precipitation on water supply; that is, stream-flow, reservoir and lake levels, and ground water levels and recharge rates. Hydrological drought and its impacts generally lag the occurrence of meteorological droughts because it takes some time

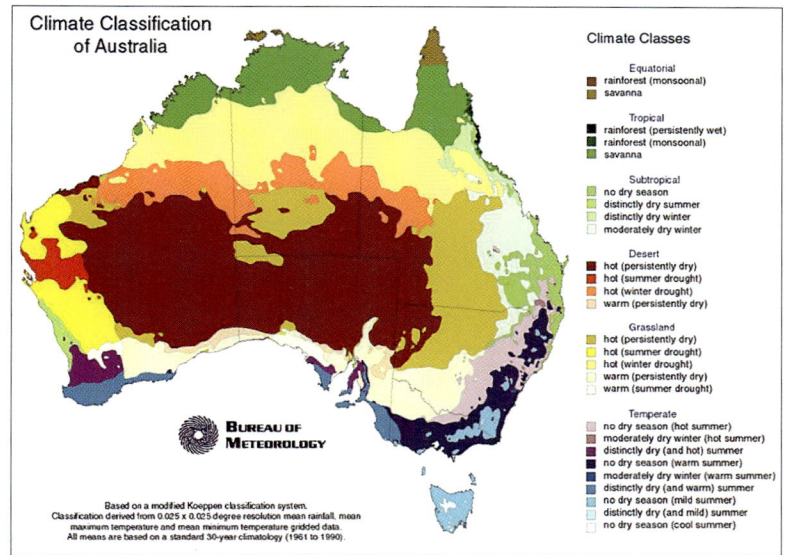

Figure 2.1 Modified Köppen climate classification of Australia

Source: Australian Bureau of Meteorology, www.bom.gov.au

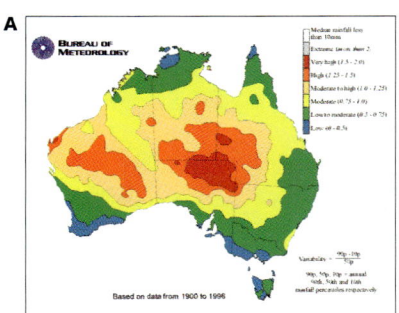

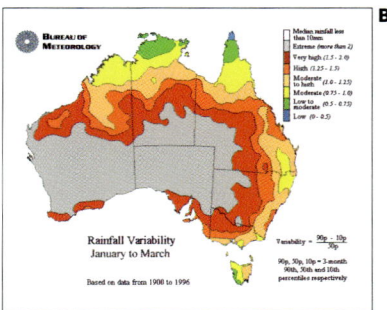

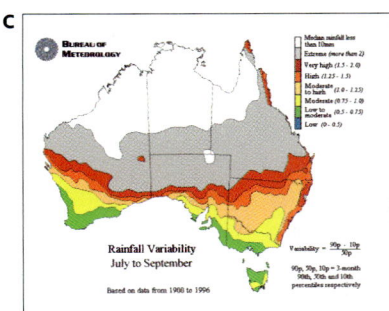

Figure 2.2 Average interannual rainfall variability over Australia: (a) annual, (b) summer (January–March), and (c) winter (July–September)

Source: Australian Bureau of Meteorology, www.bom.gov.au

33

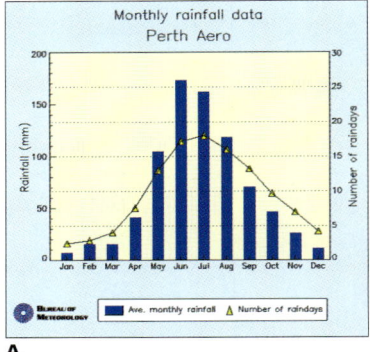
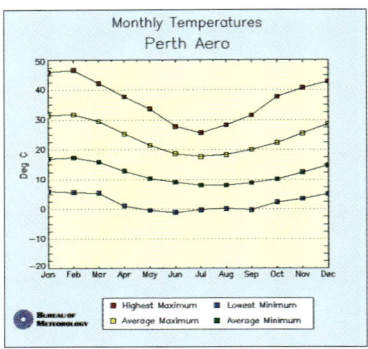

A

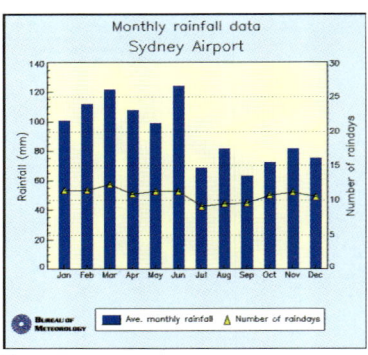
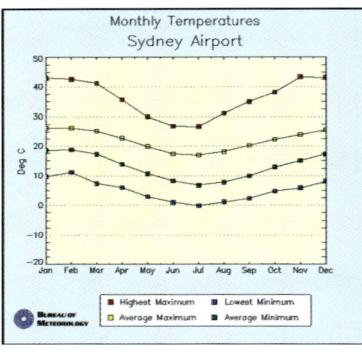

B

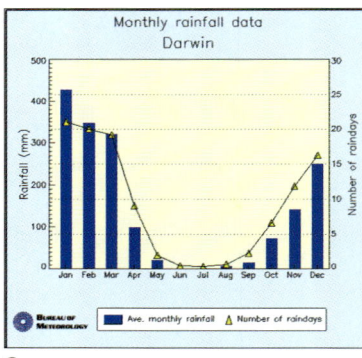
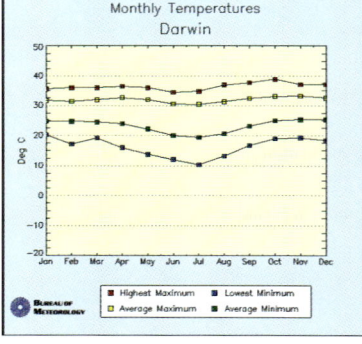

C

Figure 2.3 The seasonal cycle of rainfall and temperature at (a) Perth (subtropical west coast), (b) Sydney (subtropical east coast), and (c) Darwin (tropical)
Source: Australian Bureau of Meteorology, www.bom.gov.au

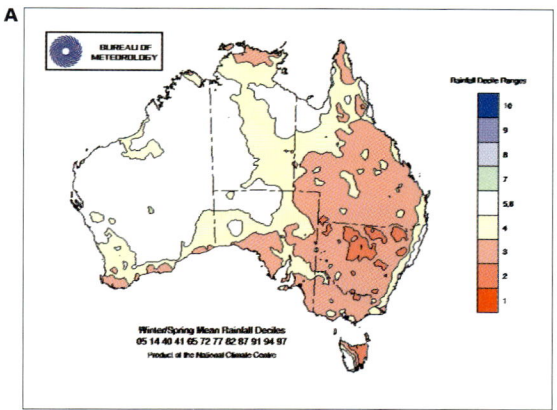

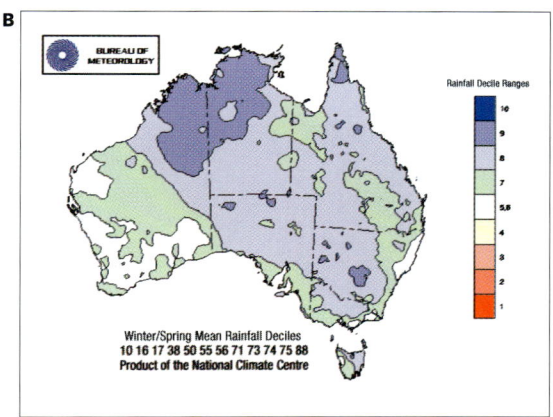

Figure 2.4 Average annual rainfall deciles for
(a) 12 El Niño years, and (b) 12 La Niña years

Source: Australian Bureau of Meteorology, www.bom.gov.au

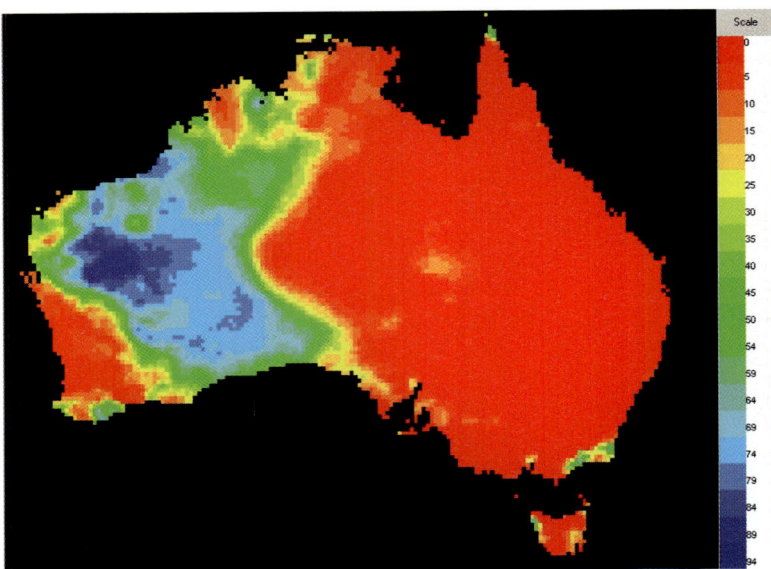

Figure 2.5 Average annual rainfall deciles for El Niño drought years:
November 1901–October 1902
Source: Bureau of Rural Sciences

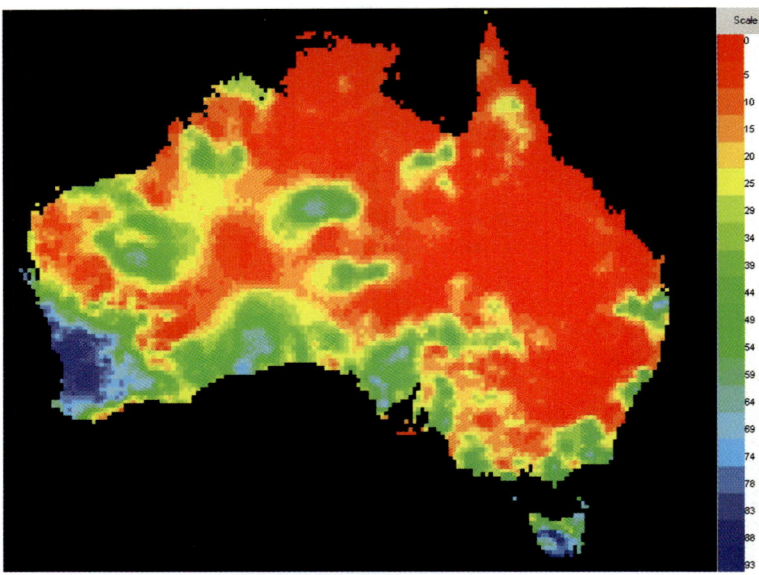

Figure 2.6 Average annual rainfall deciles for El Niño drought years:
April 1946–January 1947
Source: Bureau of Rural Sciences

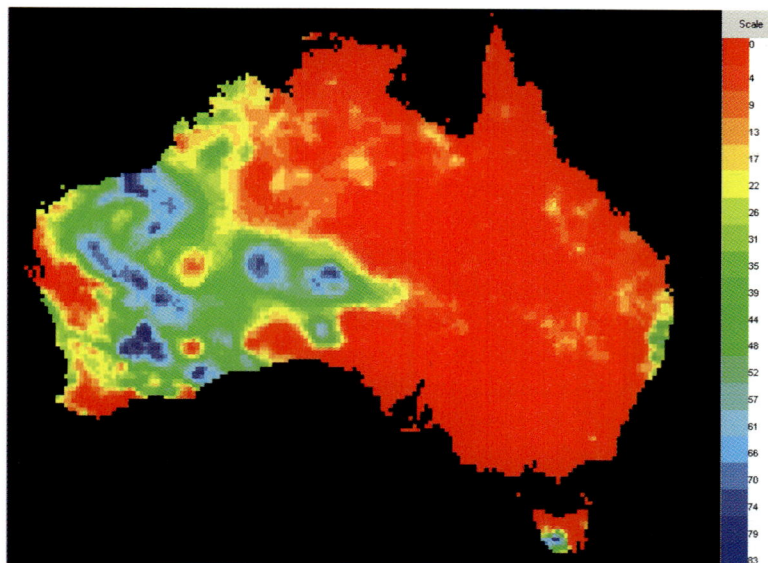

Figure 2.7 Average annual rainfall deciles for El Niño drought years:
April 1982–February 1983
Source: Bureau of Rural Sciences

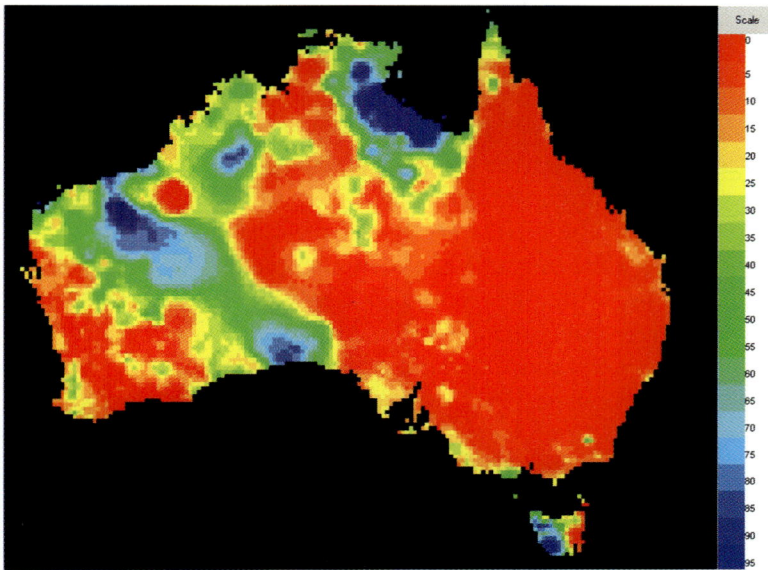

Figure 2.8 Average annual rainfall deciles for El Niño drought years:
April 2002–January 2003
Source: Bureau of Rural Sciences

for precipitation deficits to accumulate in components of the hydro-logical system, including stream-flow, ground water and reservoir levels, and soil moisture. Land-use change can influence hydro-logical drought, or even cause reductions in infiltration and run-off in the absence of meteorological drought. Thus activities such as deforestation can have effects beyond the immediate area affected by meteorological drought.

Agricultural drought associates characteristics of meteorological and hydrological drought with agricultural impacts. Factors including precipitation shortages, differences between actual and potential evapotranspiration, soil moisture deficits, reduced ground water or reservoir levels, are important in defining agricultural drought. Plant water demand depends on prevailing weather conditions, biological characteristics of the specific plant, its stage of growth, and the physical and biological properties of the soil. A good definition of agricultural drought should be able to account for the variable susceptibility of crops during different stages of crop development, from emergence to maturity.

The *sequence of impacts* associated with meteorological, agri-cultural and hydrological drought further emphasises their differences. The agricultural sector is usually the first to be affected by drought because of its dependence on stored soil moisture, which is depleted during extended dry periods. Continued precipi-tation deficiencies will affect other water users, with sub-surface water resources often the last to be affected. A short-term (three- to six-month) drought may impact only slightly on the latter, depending on the characteristics of the hydrologic system and water use requirements. When meteorological drought conditions diminish, soil water is replenished first, followed by stream-flow, surface water reservoirs and ground water. Drought impacts may diminish rapidly in the dryland agricultural sector because of its reliance on soil water, but linger for months or even years in other sectors dependent on stored surface or sub-surface supplies.

Socioeconomic drought associates the supply and demand of some water-dependent economic good, such as water supply, animal feed, fish or hydro-electric power, with elements of meteorological, hydrological and agricultural drought. It differs from these types of drought because its occurrence depends on identifying drought based on the temporal and spatial processes of supply and demand.

Socioeconomic drought occurs when water supply is unable to meet economic demand due to weather-related factors. Since both demand and supply vary with time, it is important to identify possible convergence in the trends that could signal enhanced vulnerability to drought and so increase the incidence of socio-economic drought.

Exposure and vulnerability

The short-term meteorology (days to weeks) and longer term climatology (months to years) influencing a region determine the *exposure* to drought; that is, the likelihood that atmospheric circulation anomalies leading to drought conditions will occur. Although these factors cannot be influenced by humans, some forewarning of their occurrence would allow informed planning. Considerable attention has therefore been given to developing seasonal climate forecasting.[17] In contrast to exposure to drought, *vulnerability* is determined by human activity. The impacts of drought are dependent not only on the duration, intensity and spatial extent of a drought, but also on the demands made by human activities and natural systems on available water supplies. Improved understanding of the past drought climatology of a region will provide critical information on the frequency and the intensity of historical events, and so may assist in planning for mitigating the impacts of future droughts.

A brief history of Australian drought

Given the high degree of interannual climate variability in much of Australia, it is not surprising that rainfall deficits and drought are a recurrent feature of the climate of the continent. On average a severe drought occurs somewhere in Australia approximately once in 18 years; actual return intervals vary between four and almost 40 years. The characteristics of individual droughts are just as variable, with some rainfall deficits accruing over years while other droughts are short and intense. It is also common for some regions to record good rainfall while others are in drought. Droughts in Australia affect agriculture through crop and stock losses, and are often associated with severe bushfire seasons, dust storms and soil loss, and environmental degradation in general. The most widespread and/or intense droughts, such as those of 1982–83 and 2002–03, have an immediately discernible impact on the Australian economy, and have environmental and economic consequences beyond the end of the meteorological drought.

The droughts that have made a particular impression in Australia include:

- 1895–1902
- 1914–15
- 1937–45
- 1965–68
- 1982–83
- 1991–95 and
- 2002–03.

Some of these drought periods were related to the occurrence of El Niño events (for example, 1982–83), and in some cases large parts of the country were affected (for example, 1901–02). But in other cases drought was more localised, or occurred in the absence of an El Niño event under the influence of other large-scale climatic factors. The following brief descriptions of significant drought periods in Australia during the last 100 years are based largely on material from the Australian Bureau of Meteorology *Climate of the 20th Century* project.

The Federation drought of 1895–1902 was among the most environmentally damaging types of drought, with one or two drought years following a prolonged period of generally below-average rainfall. Across much of Australia the rainfall pattern was marked by dry spells through the years prior to Federation in January 1901, particularly in 1897 and 1899. Following some rain in many areas in 1900–01 dry conditions became established across eastern Australia, with Queensland, New South Wales and Victoria in drought by mid-1902 (see Figure 2.5). The drought finally broke in December 1902. The impact on agriculture was severe, particularly on the wheat crop and on livestock; much of Queensland had been drought-affected for eight years by the end of 1902.

The next major drought was that of 1914–15, which was memorable not only for its intensity but also for the fact that much of Australia was affected, beginning with South Australia, Tasmania, Victoria and New South Wales. Unusually, both the eastern states and the south of Western Australia simultaneously suffered dry conditions that led to the total failure of the national wheat crop in 1914. There was a strong El Niño event in 1914, and severe bushfire conditions occurred in south-eastern Australia during the dry summer that year. Although this 18-month drought was not uniformly dry either in time or in space, it was the worst drought of the century in some areas; it ended in the winter/spring of 1915.

The late 1930s and 1940s were another period of generally below-average rainfall, with significant droughts occurring in 1937-38, 1940–41 and 1943–45. Severe drought conditions began in New South Wales, Queensland, Victoria and some parts of Western Australia in 1937, spreading to South Australia in 1938. Conditions in the south-west had serious impacts on the wheat crop, and in January 1939 Victoria experienced the 'Black Friday' bushfires. Following good rains during 1939, drought conditions returned in 1940 with a strong El Niño event; this was one of the driest years on record across southern Australia. Good rainfall was recorded in the second half of 1941 and in 1942, but the drought resumed in 1943 and, in many parts of Victoria, South Australia, New South Wales and Queensland, continued to mid-1945 and beyond (see Figure 2.6).

The period from 1957 to 1968 was comparable to the Federation and 1937–45 droughts in severity and areal extent. A complex succession of drier and wetter years marked this period, and the most intense rainfall deficits occurred in various parts of the country at different times, including in northern and central Australia. The eastern states were affected most during 1965–1968, with severe drought in south-eastern Australia in 1966–67, accompanied by bushfires in Tasmania and dust storms in South Australia. The drought ended late in 1968.

The 1982–83 drought was one of the most intense and widespread on record in Australia (see Figure 2.7), and has been described as having the worst overall impacts of droughts in the twentieth century.[18] This was a year in which an El Niño event developed rapidly and particularly strongly across the Pacific basin during mid-1982, and the SOI reached its lowest values in a century. Drought conditions occurred across most of Australia, but particularly in the eastern half of the country (with the exception of north-east New South Wales and south-east Queensland). A number of dust storms were generated as strong winds stripped dry, unvegetated topsoil from extensive areas in south-eastern Australia, and in February 1983 dangerous fire-weather conditions led to several large bushfires burning across South Australia and Victoria in the devastating Ash Wednesday fires of 16 February. Agricultural production was cut by an estimated 10 per cent as a result of the 1982–83 drought, with an estimated cost to Australia from all drought-related losses of around $3 billion.

The next significant droughts occurred during 1991–1995, a period characterised by the unusual (but not unprecedented) occurrence of four consecutive El Niño years. Conditions were particularly dry in Queensland, northern New South Wales, and parts of central Australia during this period. The northern wet season failed in the Northern Territory in 1991–92, and Queensland remained dry although some rain occurred further south across south-eastern Australia. By January 1994, extended drought conditions across New South Wales contributed to severe bushfires around Sydney. Although some rain fell in autumn 1994, the development of stronger El Niño conditions led to increasing drought across the country from mid-1994 into 1995; the drought that year is estimated to have reduced agricultural production by 8 per cent. This protracted drought period ended with good rains in late 1995 and 1996.

Most recently, 2002–03 saw one of the most widespread and severe droughts on record across much of Australia, this time accompanying a relatively moderate El Niño event (see Figure 2.8). In this case the Pacific Decadal Oscillation was in a negative phase (as it had been in the 1960s, for example), contributing to the magnitude of the ENSO impact in Australia. Severe rainfall deficiencies in almost every state were exacerbated by well above average temperatures, particularly in parts of Queensland and New South Wales. Maximum temperatures averaged 1.59°C above normal in 2002, and were 1.65°C higher than average in March–November 2002 (a record for Australia).[19] Drought conditions from April 2002 to January 2003 resulted in a severe bushfire season in areas of Victoria, Tasmania, New South Wales and the Australian Capital Territory, with the Canberra fires of 18 January 2003 the second most costly in Australia's history.

Drought has been a recurrent feature of the Australian environment throughout the period of instrumental record; that is, since the mid to late nineteenth century. Its occurrence has been variable in both space and time. However, an increased understanding of the basic climate state and the nature of climate variability, and of climatological factors influencing drought occurrence and severity (for example, ENSO and the Pacific Decadal Oscillation), has allowed the development of seasonal climate outlooks and advisories that are useful for natural resource and agricultural planning and management.[20] An additional factor that must be taken into account, however, is the possibility of climate change.

Climate change and the future

Against the background of a climate system characterised by a high degree of variability on a range of time scales, it is now acknowledged that some degree of human-induced change in climate is occurring and will continue to occur into the future.[21] The changes that have been observed over the last 100 years or so, particularly in air temperature and, in more complex ways, in rainfall, are due to a combination of natural variability and increasing concentrations of carbon dioxide and other heat-absorbing gases in the atmosphere. The amounts of these gases present in the atmosphere have risen exponentially since the 1850s, largely as a result of increasing combustion of fossil fuels, agricultural practices, and land-use change (including deforestation). The result of increasing atmospheric concentrations of the greenhouse gases is a general increase in near-surface air temperature. Over the same period, industrial processes have led to an increase in the amount of sulfate particles in the atmosphere, which have a cooling effect on near-surface air temperatures and so moderate greenhouse warming to some extent.[22]

The overall response in the climate system has been a globally averaged warming of approximately 0.6°C since 1900. The observed changes in global air temperature and carbon dioxide in this period exceed the maxima reached during at least the past 500 000 years.[23] In Australia the average temperature across the continent has risen by 0.7°C between 1910 and 1999, with much of the warming occurring in the second half of the twentieth century.

The warmest year on record across the continent was 1998. As has occurred in many other regions, Australian night-time minimum temperatures have increased more rapidly than daytime maximum temperatures, leading to a reduction in the difference between daytime and night-time temperatures (the diurnal temperature range). Over the same period, rainfall, which is inherently more variable in space and time than temperature, has not shown any clear trend in the continental average. There are, however, some identifiable trends in regional rainfall.[24]

A significant international scientific focus on possible future greenhouse gas emissions scenarios[25] has resulted in projections of global average warming of between 1.4 and 5.8°C by 2100 (relative to 1990), with an average warming rate of between 0.1 and 0.5°C per decade. Simulations of the climate system using complex Global Climate Models and these global warming projections allow the development of scenarios for average temperature and rainfall change across Australia in the future.[26] Regionally specific projections are being developed using finer resolution (limited area) models such as the CSIRO Atmospheric Research DARLAM model.[27] It is not possible to predict conditions in individual years, since natural climatic variability will continue to affect the inter-annual fluctuations of climate in unpredictable ways.

Overall, annual average temperatures are expected to rise by between 0.4 and 2.0°C across much of Australia by 2030, with slightly smaller increases in the south and larger increases in the north-west. By 2070, temperatures are expected to be between 1.0 and 6.0°C higher than in 1990. The largest increases occur in summer, and a greater frequency of very hot summer days (with temperatures exceeding 35°C) and a concomitant decrease in the frequency of winter days with sub-zero temperatures are predicted across the continent.

The projected changes in rainfall across Australia are more spatially variable, and carry higher uncertainties than do the temperature projections.[28] Annual rainfall changes of between –20% and +20% are predicted by 2030, with the biggest reductions in the south-west and parts of the south-east, but little change in the tropics across the northern parts of the continent. By 2070, rainfall may have changed by as much as 60 per cent in some areas. Seasonal patterns of rainfall change are complex: in winter and spring the trend is towards drier conditions, whereas in summer and autumn some areas become wetter, while others are drier. A further important aspect of possible future rainfall conditions is the predicted increased frequency of extreme rainfall events.[29]

Future scenarios for Australian climate in the twenty-first century are thus characterised by higher temperatures and, in some areas, by drier conditions, particularly in the southern and western parts of the continent. There is, however, marked seasonal variation in the rainfall projections. Research on ENSO behaviour under greenhouse warming scenarios indicates that there is unlikely to be any significant change in ENSO; that is, the interannual fluctuations of sea surface temperature and atmospheric circulation will continue to produce conditions that will result in El Niño and La Niña events in future. However, these events will be impacting on an environment potentially more vulnerable to extreme conditions due to the changes in climate described here. It seems probable that these changes, with adjustments in both the long-term mean and in overall variability, will result in a general shift towards increased frequency of high-temperature, extreme-rainfall events. Human-induced climate change has been suggested as the cause of the combination of severe rainfall deficits and high temperatures (with associated soil moisture loss and vegetation drying) that charac-terised the 2002–03 drought across large parts of Australia.[30] The possibility of such drought impacts recurring in future raises impli-cations for land and water resource management and for agriculture that will have to be addressed in policy, planning and management.

Summary and conclusions

In this chapter I have established the climatological context for the discussion of drought in Australia. The climate of the continent is characterised by clearly defined rainfall and temperature zones, influenced by its location between the tropics and the midlatitudes and by the large east-west extent of the landmass. The most significant features of the climate of Australia are the strong seasonality of rainfall, the degree of change in seasonality during the last 100 years or more, and the relatively large interannual rainfall variability. Year-to-year fluctuations in climate are due to often-complex combinations of large-scale influences, including the

El Niño Southern Oscillation, the Pacific Decadal Oscillation, the Indian Ocean Dipole, and others. The result of the nature of the Australian climate is that conditions of rainfall deficit or drought are relatively common in many parts of the country.

Defining drought is not a simple matter, however, and a range of definitions has been considered. Different components of human and natural systems respond to a rainfall deficit in different ways, often complicating the assessment of the beginning and end of drought conditions. There is no doubt, however, about the impacts on Australia of the most severe droughts of the last century; these are evident in an overview of the drought history of the continent. The importance of using our understanding of such past climatic variability, and drought in particular, to inform both current decision-making and planning for the future is highlighted in a discussion of scenarios for possible future climate change in Australia. If rising global temperatures and changing climate may result in an increased frequency of high-temperature and extreme-rainfall seasons, the implications for land and natural resource management should inform policy formulation and management planning.

---

1 Groves 1994.

2 Allan and Lindesay 1998.

3 Blench and Marriage 1999, p. 9.

4 For example, Colls and Whitaker 1990; Linacre and Geerts 1997; Sturman and Tapper 1996.

5 For example, Colls and Whitaker 1990; Hobbs 1998; Sturman and Tapper 1996.

6 Lindesay and Johnson 2003.

7 For example, Allan *et al.* 1996; Hobbs *et al.* 1998.

8 Allan *et al.* 1996.

9 Allan *et al.* 1996.

10 Allan *et al.* 1996.

11 Power *et al.* 1999.

12 Saji *et al.* 1999.

13 White, WB 2000.

14 Power *et al.* 1999.

15 For example, Glantz, MH 2000; White *et al.* 1997.

16 Wilhite and Glantz 1987.

17 McKeon *et al.* 1993; White *et al.* 1999.

18 Allan and Heathcote 1987; Allan *et al.* 1996; Glantz *et al.* 1987.

19 Karoly *et al.* 2003.

20 Nicholls 1997a; White *et al.* 1999.

21 Houghton *et al.* 2001.

22 Houghton 1994; Houghton *et al.* 2001.

23 IPCC 2001.

24 Bureau of Meteorology, *Climate of the 20th Century*.

25 Houghton *et al.* 2001.

26 CSIRO 2001.

27 For example, Whetton *et al.* 20010.

28 CSIRO 2001.

29 CSIRO 2001.

30 Karoly *et al.* 2003.

3

Government responses to drought in Australia

Linda Courtenay Botterill

After 1788, the Australian landscape was gradually transformed as farmers introduced European agricultural practices in order to secure a food supply for the new colony. Attempts to do so met with limited success and as early as October 1788 Governor Phillip was writing to London urging the Government to provide free settlers with agricultural skills as food shortages were a real concern.[1] The first Europeans to engage in agriculture were soldiers and emancipists who were given plots of land. They had little knowledge of agriculture and even less of Australian conditions. An early report on the state of agriculture and trade in the colony of New South Wales stated that the 'uncertain climate' was 'not generally favourable to the growth of European grains'[2] and suggested that the future of the colony:

> ... will be that of pasture rather than tillage, and the purchase of land will be made with a view to the maintenance of large flocks of fine-woolled sheep; the richer lands, which will generally be found on the banks of the rivers, being devoted to the production of corn, maize and vegetables.[3]

The colonies experienced severe droughts in the 1820s and 1895–1902,[4] the latter coinciding with general economic depression.

Drought was considered to be a natural disaster and, although responsibility for responding to natural disasters lay with the State governments, the Commonwealth gradually became involved until it was contributing substantial sums to drought relief through natural disaster relief arrangements with the states and on an *ad hoc* basis through special purpose legislation. In 1989, drought was removed from the Commonwealth–State natural disaster relief arrangements, and this decision was followed in 1992 by the announcement of a National Drought Policy.

A brief account of recent drought policy developments[5]

Whether drought is considered to be a natural disaster, as it was prior to April 1989, or as a farm management issue, which it has been since that time, there is a strong case to be made that drought policy is a state responsibility. Section 51 of the Constitution, which sets out the areas in which the Commonwealth can legislate, does not include agriculture or natural disasters and as such these issues fall within the legislative responsibilities of the states. The State and Territory governments implement their own drought policies and there is a range of approaches, both to the declaration of droughts and the nature of the policy response. Nevertheless, the Commonwealth government has been involved in responding to drought to varying degrees since Federation, and since 1992 a concerted effort has been made by State and Commonwealth governments to develop a national policy response.

Rural policy has benefited from longstanding consultative arrangements between the Commonwealth and State governments through a Ministerial Council and its associated Standing Committee of officials. Originally established as the Australian Agriculture Council in 1934, the Ministerial Council is the oldest of the Commonwealth-State consultative arrangements and has been described as very cooperative, if not the most successful.[6] Since the early 1990s the Council has been particularly active on the issue of drought policy and it is these recent developments that are the focus of the brief historical summary that follows. It is important to note that alongside the policies described below, the various State and Territory governments continue to implement their own drought relief programs; however, these are beyond the scope of this chapter, which focuses on the attempts to develop a cohesive national response to drought.

As noted above, until 1989 drought was considered to be a natural disaster, along with cyclones, floods, bushfires, earthquakes and other sudden, unpredictable natural occurrences or 'Acts of God'. The Commonwealth first became involved in disaster relief in 1939, when it provided £1000 to the government of Tasmania for bushfire relief, and there was also a number of occasions on which special purpose legislation was passed through the Commonwealth Parliament to assist the states in their response to natural disasters. In the case of drought, examples include the States Grants (Drought Assistance) Acts of the 1960s.

In 1971 the Commonwealth government changed the basis on which funding was provided to the states for disaster relief when it imposed a threshold on the states below which the Commonwealth would not contribute. Prior to this decision, Commonwealth support had matched state expenditure on a dollar-for-dollar basis for personal hardship relief and the restoration of public assets.[7] A similar change to the basis for Commonwealth involvement was made in 1978 and again in the 1980s when the formulas for determining these thresholds were amended.

In April 1989, the Commonwealth government announced that drought would no longer be covered by the Commonwealth–State arrangements. There were budgetary and political reasons behind this decision; however, it was also becoming increasingly untenable to argue that drought was a natural disaster in a country with a climate as variable as Australia's. The decision to cease treating drought as a natural disaster was endorsed by a Drought Policy Review Task Force, which reported in May 1990[8] and a Senate inquiry, which reported in 1992.[9] In policy terms this could be considered a 'fundamental decision',[10] as drought was redefined from being a disaster to a business risk to be managed by farmers like any other risk; for example, currency or interest rate movements. Although the decision to remove drought from the natural disaster relief arrangements marked a policy watershed, it is worth noting that attitudes towards drought among farmers and across the broader community did not necessarily shift quickly to the new 'risk management' paradigm.

After months of negotiations throughout 1991 and 1992, the Commonwealth and State Ministers announced in July 1992 that they had agreed on a National Drought Policy. The objectives of the new policy were to:

- encourage primary producers and other sections of rural Australia to adopt self-reliant approaches to managing for climate variability;
- facilitate the maintenance and protection of Australia's agricultural and environmental resource base during periods of climate stress; and
- facilitate the early recovery of agricultural and rural industries, consistent with long-term sustainable levels.[11]

Support was available to help farmers adapt to this new risk management environment through the revised Rural Adjustment Scheme, which commenced on 1 January 1993. The scheme offered support for farmers in two forms. Under what became known as the 'normal' provisions of the scheme, farmers could obtain grants and interest rate subsidies of up to 50 per cent to improve their farm operations and their management skills. In recognition that events occur for which the best manager could not be expected to prepare, including but not exclusively drought, the scheme included 'exceptional circumstances' provisions. In an exceptional circumstance, an eligible farmer could receive assistance in the form of interest rate subsidies of up to 100 per cent. The term was not defined in the legislation or the second reading speech, and determination of what constituted an exceptional circumstance quickly came to dominate debate over drought policy.

Unfortunately, the National Drought Policy was announced as Queensland and New South Wales were feeling the effects of the beginning of one of the worst droughts of the twentieth century. This meant that the exceptional circumstances provisions appeared to be anything but, in terms of the frequency of their use. From the first day the new legislation came into force the exceptional circumstances provisions were activated, although ironically the first declaration was for excessive rain in South Australia and Victoria. From that point, however, more and more areas were declared to be in exceptional circumstances and debate continued over definitions of drought and the implementation of the scheme.

As noted, the primary mechanism for responding to drought under the new National Drought Policy was the exceptional circumstances provisions of the Rural Adjustment Scheme that, by its very nature, was targeted at facilitating structural adjustment in agriculture. Structural adjustment pressures affect all industries in a developing economy. Just as vinyl record producers responded to market developments by moving into the production of compact discs, farmers face pressure to adjust their production in response to changing market conditions for their products. In recent years, for example, this has seen shifts in production from wheat of fair average quality to the production of high-protein durum wheat, for pasta manufacture, and other specialty wheats. As well as changing consumer demand, Australian farmers have also experienced ongoing declining farm terms of trade, which have required productivity improvements and increased management skills. To support this process, the Australian government has been offering assistance to farmers faced with structural adjustment pressures since the introduction of the *Loan (Farmers' Debt Adjustment) Act 1935*. In recent years this support has been focused on improving the prospects for farmers who are viable in the long term, and providing exit assistance in the form of re-establishment grants and income support to those who are not.

Using the Rural Adjustment Scheme as a mechanism for supporting drought-affected farmers was therefore limited: it only assisted those who were eligible for support under the scheme; that is, farmers who had good prospects of long-term profitability. As the1990s drought worsened, it became apparent that there were farm families experiencing genuine hardship who were ineligible for exceptional circumstances support and who were also unable to access the general social welfare safety net for a variety of reasons. Although the issue of farmers' access to social welfare is beyond the scope of this chapter, it is worth noting that the general social welfare safety net is primarily targeted at wage and salary earners and does not cope well with the income support needs of the asset-rich or the self-employed. Many farmers who were income-poor during the drought in the 1990s found themselves excluded from social security payments due to the assets test.[12] This included farmers' spouses who were available and willing to work, but unable to access the unemployment benefit due to the value of the family farm.

Following a visit to south-west Queensland in September 1994 by the Prime Minister, Paul Keating, the Commonwealth government announced that it would supplement the support available through the Rural Adjustment Scheme with an income support payment for farmers in exceptional circumstances drought areas. The Drought Relief Payment was not limited to farmers with a long-term viable future in agriculture, but was only available to those within areas declared to be in exceptional circumstances drought.

In 1997, following the change of government at the Commonwealth level, the Rural Adjustment Scheme was scrapped; however, the exceptional circumstances concept was retained and the Drought Relief Payment became the Exceptional Circumstances Relief Payment. As has been the case throughout the existence of the National Drought Policy, the Ministerial Council continued to debate definitional and funding issues surrounding the exceptional circumstances program. The current (2003) definition of an exceptional circumstance is that:

- The event must be *rare* and *severe*.
- The effects of the event must result in a *severe* downturn in farm income over a prolonged period.
- The event must not be predictable or part of a process of structural adjustment.

The criteria also specify that 'The key indicator is a severe income downturn, which should be tied to a specific rare and severe event, beyond normal risk management strategies employed by responsible farmers ... The severe downturn should be for a prolonged period and of a significant scale.' 'Rare' is taken to be an event that occurs 'on average *once in every 20 to 25 years*', but can also include 'a combination of exceptional factors formulating an event'.[13]

One of the biggest problems with the exceptional circumstances program since its inception has been the need to define geographically those areas that qualify for support. This has been labelled the 'lines on maps' problem as inevitably any such definition excludes some needy farmers whose situations may be indistinguishable from those of their neighbours on the 'right' side of the lines. To ameliorate this issue, Ministers agreed in August 2001 that:

Farmers outside the defined zone, but who are in reasonable proximity and can also demonstrate that they are affected by the same exceptional events, will be eligible to make application under the same terms and conditions as those within the defined zone.[14]

This decision to introduce 'buffer zones' has simply blurred the lines rather than addressing the inequity. More recent events have seen a further blurring of the exceptional circumstances concept with the announcement by the Commonwealth Minister for Agriculture, Fisheries and Forestry, Warren Truss, that income support payments would be available to farmers on the basis of a *prima facie* case that they qualified for support. If the application for an exceptional circumstances declaration is subsequently rejected, farmers will still be able to receive up to six months of welfare support at the equivalent of the unemployment benefit. Successful applications will result in income support payments for two years[15] and business support.

Priorities, trade-offs and ideological gravity

Any account of drought policy that focuses on the series of government decisions, announcements and the implementation of new programs will invariably imply the existence of a rational, linear process by which policy develops logically. This is a misleading impression. The clear-cut policy announcements hide the negotiations and compromises between the key players, in this case the members of the Ministerial Council, and can understate their importance. For example, the Drought Relief Payment was not the result of the standard Departmental advice process, but was initiated by the Minister's Office.

Policy is an ongoing process of consideration and reconsideration of ideas, options and instruments. As Braybrooke and Lindblom have noted, 'policy-making proceeds through a sequences of approximations. A policy is directed at a problem; it is tried, altered, tried in its altered form, tried again, and so forth'.[16] As this process evolves and policies are developed incrementally, opportunities arise for policy learning to occur. In other words, each new iteration allows policy-makers to assess the performance of existing measures and redress perceived deficiencies. In the long run, policy learning should result in better policy, although as May notes, 'this does not necessarily entail discovery of an 'objective' truth about a policy, instrument, problem or goal'.[17] May has described two types of policy learning: instrumental policy learning and social policy learning. '*Instrumental policy learning* entails lessons about the viability of policy instruments or implementation decisions. *Social learning* entails lessons about the social construction of policy problems, the scope of policy or policy goals'.[18]

Both types of policy learning are evident in the development of drought policy since 1989. In terms of instrumental policy learning, the frequent revisiting of the definition of exceptional circumstances suggests policy-makers recognised the limitations of existing approaches and were seeking to improve on earlier attempts at identifying areas that would be eligible for support. The use of geographical boundaries to define eligibility has caused persistent problems for policy-makers. Recent attempts to blur the boundaries by allowing limited applications for support from those outside exceptional circumstances areas is a recognition of this problem; however, to date the government has not proposed an alternative approach that addresses the problem completely. As the buffer zones are applied, it is likely that policy-makers will again identify limitations to the approach and revisit the issue of geographical boundaries.

In 1994, it became clear that the provision of exceptional circumstances support exclusively by way of the interest rate subsidies available through the Rural Adjustment Scheme was inadequate in alleviating the hardship being experienced by many farmers. The policy instrument that had been adopted to offer support was found to be lacking, and alternative measures were sought, resulting in the development of an additional program, the Drought Relief Payment. This development was a recognition that business support alone was inadequate and a welfare program was also necessary to meet farmers' needs.

The rejection by the rural policy community of the construction of drought as a natural disaster is an example of social policy learning. Although the reasons for removing drought from the natural disaster relief arrangements were varied, the decision was followed by a growing consensus that drought in Australia needed to be viewed from a different perspective. It should be noted that the new construction of drought as a business risk was generally accepted in the policy community; however, it is unclear whether a major shift has occurred in the way farmers view drought. Also, it is clear from the way major media organisers have promoted the 2002 Farmhand appeal that they still see drought in disaster terms and expect to evoke a similar response from potential donors to their appeal. These issues are explored further in this volume in Hayman and Cox's chapter on farmers and risk and Wahlquist's discussion of the media, the public and drought.

The fluid policy process described above has been variously characterised as a 'garbage can'[19] or 'primeval soup'[20] out of which policy outcomes emerge. This incremental policy development process allows for fine-tuning of policy approaches; however, the question of the policy endpoint can be overlooked. In the area of drought policy, this volume proposes an end point in which we learn to 'be Australian' in our approach to climate. This end point then indicates the direction in which policy needs to move, rather than relying on incremental change that may or may not lead us in this direction. Our approach does not reject incrementalism, but suggests that we need to steer the process in the direction of our goal. The current National Drought Policy contains such a goal; however, recent policy developments such as the provision of support on the basis of *prima facie* evidence, appear to have digressed from the objective of self-reliance and risk management. Policy-making during drought creates an impression that political pressure will result in policy change and support for farmers. This expectation of government support in response to lobbying during drought can detract from the objectives of the National Drought Policy.

The goal-setting process, as well as the development of detailed policy once those goals are agreed to, does not happen in a vacuum. As we argue in this volume and as May implies, there is no such thing as a single, objectively correct policy response to complex issues. No policy solution can address all of the social, economic and environmental issues that are relevant to the problem at hand. A range of policy options can be developed, each of which contains trade-offs between competing values that are either explicit or implicit. The introduction of the Drought Relief Payment, for example, represented a trade-off between the economic objective of promoting structural adjustment in agriculture, by avoiding the provision of *de facto* subsidies to otherwise unviable businesses, and social justice concerns about the ability of farmers in drought-affected areas to feed and clothe their families. An important distinction between the Drought Relief Payment and the support available through the Rural Adjustment Scheme was that the latter was a budget-limited program, which meant that support was rationed. The Drought Relief Payment and its successor, the Exceptional Circumstances Relief Payment, by contrast, were entitlement-based. All farmers who qualified for support received the full amount to which they were entitled.

To illustrate the range of options available to policy-makers and how different priorities combine to produce different outcomes, the following typology is proposed, which is laid out in Figure 3.1. Each of the four quadrants represents or contains policy approaches that have struck different balances between competing values or reflect different sociopolitical contexts in which the policy was developed. On the vertical axis are indicated varying levels of support for the farm business and on the horizontal, support targeted at the welfare needs of the farm family. Each axis represents a continuum with the outer ends indicating more extreme positions. As well as illustrating the balancing act facing policy-makers, the figure is useful in illustrating how policies can shift over time in emphasis—effectively moving between quadrants.

The four quadrants, moving clockwise from the top left, represent the following policy models.

The *farm business model* is characterised by an emphasis on support for the farm business. This model is based on assumptions about farming that are anchored in the free market model, but see sufficient market failure to justify varying levels of government intervention to help farm businesses achieve their full potential. Examples of the type of support included in this quadrant are the interest rate subsidies offered under the 1992 Rural Adjustment Scheme, and long-term, low interest loans or other government financial support aimed at assisting the farm business in response to perceived failures in the commercial financial market. The 1992 Rural Adjustment Scheme offered enhanced interest rate subsidies for farmers experiencing exceptional drought, but only if they qualified for support through the scheme under normal circumstances; that is, they had long-term prospects of profitability. This quadrant would also contain income-smoothing mechanisms like farm management bonds or revenue contingent loans,[21] as well as programs aimed at skills development. Income-smoothing, interest rate subsidies for farm improvement and to tide people over during drought, skills development and planning subsidies were all offered on the assumption that with better management, farm businesses could be drought proofed for all but exceptional events. The further north the intervention is located in this quadrant, the less likely that the intervention is tied to structural adjustment objectives.

The *agrarian model* encompasses support to both the farm business and the farm family. This model is influenced by agrarian sentiments, which attribute to farming a range of characteristics that go beyond the economic. In Australia this view has been described as countrymindedness[22] and is arguably part of the national self-image that extends beyond the rural to influence the values of urban-based Australians. To a large extent, the agrarian model could be seen as the most politically responsive of those described in the policy map. The agricultural policies of the 1950s and 1960s in Australia would fit in this quadrant, as they provided high levels of intervention through statutory marketing arrangements, home price schemes and so on, coupled with an emphasis on the importance of supporting farm incomes. This model includes income support that is decoupled from farm business objectives, either partially or completely. Some of the assumptions underpinning this model include the idea of the importance of Australian agriculture *per se* and the appropriateness of government intervention to protect the sector and the people involved in it.

The *welfare model* suggests an emphasis on the welfare needs of the farm family with less focus on supporting the farm business. Recent changes to the exceptional circumstances relief payment place that scheme in this quadrant as illustrated, as does the phasing out of interest rate subsidies that reduces the amount of business support on offer. This model can be seen as consistent with the free market model; there is reduced intervention in agricultural business, but with a recognition that farm families have difficulty in accessing the general social security safety net and may need specially targeted assistance packages that address their needs in times of financial difficulty.

The *free market model* relies on market forces to shape agriculture. In contrast to the agrarian model, this is the most 'rational' of the policy approaches—adopting a cohesive and consistent approach to policy based on clearly understood assumptions about the operation of markets and the motivations of economic agents. Policies on the right of this quadrant would include farm exit programs designed to encourage the reallocation of farm resources to more efficient operators, while those towards the bottom edge would see little intervention in agriculture.

Some examples of recent drought policy instruments have been placed in their relevant quadrants as illustrations. Policy-makers are restricted in the areas of the policy space in which they can operate due to the existence in any particular political context of a 'centre of ideological gravity' that 'pulls' policy options in its direction. They are also prevented from going to the extremes of the policy map by the pressure and influence applied by competing stakeholders.

For example, the Drought Relief Payment of 1994 constituted a movement in policy direction from the farm business to the agrarian model. This movement was generated in response to political pressure in the form of the original 1994 Farmhand Appeal, intense sympathetic media coverage, increased pressure from the opposition and effective lobbying of the Minister. However, the prevailing neo-liberal approach to economic policy constrained the policy to the extent that the payment was only available in regions declared to be in exceptional circumstances under the provisions of the Rural Adjustment Scheme. Also, an assets test was applied to off-farm assets. The centre of ideological gravity in this case was located somewhere in the free market quadrant and acted as a counter to agrarian impulses which might have dictated that the payment be available to all farmers experiencing drought of any magnitude.

The model also illustrates the non-linear nature of drought policy. As policy has changed, the mix of values has shifted. To some this is seen as 'back-sliding', as approaches are seen to be compromised in order to meet competing objectives. In terms of structural adjustment objectives, the introduction of the Drought

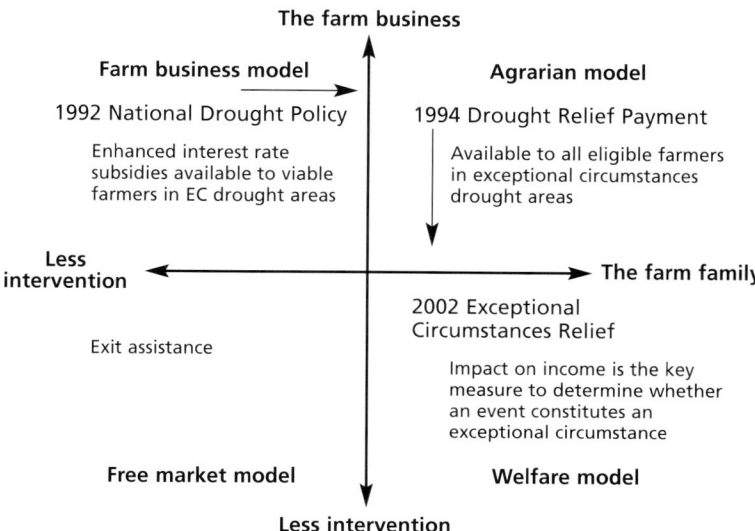

Figure 3.1 Rural policy map: drought policy

Relief Payment constituted a softening of the approach that only saw government support, other than exit assistance, offered to farmers with long-term viable futures in agriculture. The move from the farm business model to the agrarian model reflected a move in the relative importance of the farm family and the farm business in the eyes of decision-makers. An alternative view of this shift would be that social learning has occurred as policy-makers have reconstructed drought policy with a greater focus on its welfare impact than the original 1992 National Drought Policy, which clearly framed climate variability as a farm management issue.

Responding to drought in Australia

Drought policy in Australia is predominantly about farm policy. In recent years the policy approach has emphasised the importance of farm management practices to the capacity of a business to withstand drought. Contemporary policy approaches emphasise self-reliance, risk management and drought preparedness, objectives that are generally supported by the research reported in this volume. These approaches to drought reflect the nature of the Australian environment; they have moved away from a disaster response; they are more likely to encourage sound natural resource management practices and reward good management. However, the challenge for policy-makers is to promote sound management practices in the face of the emotionally charged policy environment in which such policy is so often developed, while responding sympathetically to the hardship being experienced. Policy-makers seeking to achieve structural adjustment in agriculture are understandably concerned that welfare programs developed in response to farm poverty not undermine their objectives. This may be one of the hardest issues facing rural policy-makers. As Schapper argued over 30 years ago:

> At the one and the same time in Australia, there is need for efficient farming and there is concern for inadequate income farmers. But there is no political or economic mechanism which automatically ensures harmony between efficiency and welfare. This can be resolved only by government policy.[23]

Juggling the trade-offs between these objectives in order to achieve broadly acceptable and effective policy is not an easy undertaking. Like drought, policy is context specific. The prevailing economic, environmental, political and social conditions influence the goals of the policy and the types of policy instruments that are acceptable to the community. This means that the development of drought policy is ongoing and in order for it to be successful, policy learning needs to occur.

Since 1989, drought policy in Australia has developed incrementally, based on the principles originally set out by the Drought Policy Review Task Force. Remaining unchallenged over the period since the Drought Policy Review Task Force reported in 1990 is the principle that normal drought should be managed like any other risk related to the farm business. Programs have been introduced to provide farmers with the tools and training to improve their risk management skills. Also unchallenged is the concept of restricting government support to those experiencing 'severe' or 'exceptional' drought. Defining these events has proved problematic and policy-makers and stakeholders still grapple with distinguishing the events that can be encompassed within normal risk management and those that are so unusual as to justify government intervention.

The policy-making process is complicated by Australia's federal system of government. As noted, constitutionally, much agricultural policy is the responsibility of the State governments, which have their own constituencies, political agendas and objectives to meet. The Ministerial Council, currently the Primary Industries Ministerial Council, has provided a useful, and mostly cooperative, forum for the discussion of drought policy between the different levels of government. The National Drought Policy was agreed through the Council and amendments to drought support arrangements have been a regular agenda item for Ministers' consideration. Nevertheless, the federal system can limit the effectiveness of policy approaches and introduce contradictions. For example, the 1992 drought policy called for the phasing out of transaction-based subsidies to farmers during drought, largely due to concerns about their adverse environmental impacts. In spite of this, as recently as August 2002 the NSW Premier was announcing the payment of such subsidies. The pattern of State and Commonwealth elections means that at any one time, there is a strong likelihood that one of the participants in Ministerial Council deliberations is facing an imminent poll.

The structure of the exceptional circumstances declaration process has also been difficult. With the decision to declare, and most of the funding responsibility resting with the Commonwealth Minister, there has been little incentive for State Ministers to ensure that applications for assistance are thoroughly vetted at State level before proceeding to the Commonwealth government for a decision. This has particularly been the case where the State and Commonwealth Ministers have represented different political parties and there have been opportunities for political sniping about levels of responsiveness to the plight of drought-affected farmers. In August 1999 this issue was addressed by the Ministerial Council with Ministers agreeing that 'future Exception [sic] Circumstances (EC) applications will not be submitted to the Commonwealth unless the cases can be fully documented and the State or Territory Minister (or peak industry body) is reasonably confident that the case fully meets EC guidelines'.[24]

Similarly challenging has been agreeing on the appropriate funding arrangements between the Commonwealth and State governments. The 1992 Rural Adjustment Scheme was funded 90:10 by the Commonwealth and State governments for interest rate subsidies up to 50 per cent, which were available under the 'normal' provisions of the scheme. When up to 100 per cent subsidies became available during exceptional circumstances, the subsidy above 50 per cent was funded 50:50 by the Commonwealth and the States. The Drought Relief Payment was funded entirely by the Commonwealth and, as noted, unlike the Rural Adjustment Scheme that was budget limited and therefore involved a degree of rationing, the Drought Relief Payment was entitlement-based, with all eligible applicants receiving the support for which they qualified. Under the Agriculture Advancing Australia package that replaced the Rural Adjustment Scheme, the component known as 'FarmBis', aimed at improving risk management, was funded 50:50 by the Commonwealth and the States. In March 2001, the issue of the appropriate arrangements for joint funding of business support measures was raised at the Ministerial Council and officials were requested to report back at the August 2001 meeting. The funding issue was not resolved at the meeting and it was again on the agenda for the May 2002 Ministerial Council. At the time of writing, political bickering about the funding arrangements continues between the Commonwealth Minister and his State counterparts.[25]

With the 1989 decision to remove drought from the natural disaster relief arrangements, the Commonwealth set the framework for a policy approach that recognised the reality of the Australian climate. The incorporation of the concept of severe drought into this framework has, however, complicated the policy process. Defining exceptional events is an ongoing challenge for policy-makers, as is constructing appropriate policy responses that address the very real needs of drought-affected farm families without detracting from the goals of increased self-reliance, improved risk management and sound natural resource management. Striking the balance between social, environmental and economic concerns will inevitably reflect the policy priorities of the Commonwealth, State and Territory governments of the day. The drought policy process has been a story of disjointed incrementalism[26] with each drought seeing revision of the policy settings in the face of political pressure. In the drought of the 1990s, the National Drought Policy was amended through the introduction of the Drought Relief Payment. We have seen similar reactions in the 2002 drought, with the introduction of buffer zones and the provision of support to farmers with a *prima facie* case for support.

All of these developments could be seen as 'back-sliding' on the policy objective of self-reliance and risk management. Alternatively, they could be regarded as a recognition that the balance between conflicting objectives needs to be constantly fine-tuned as part of a process of policy-learning.

The incremental nature of the policy process and the ongoing review of drought policy highlights that there is no single, definitive drought policy approach. In order to ensure that the incremental process produces progressively *better* drought policy, policy-makers can seek the most up-to-date research on drought to enhance their search for policy approaches. The following chapters set out a range of different perspectives on drought and drought policy—all of which have important implications for the development of better drought policy in Australia.

---

1 Clark 1950, p. 63.

2 Bigge 1966, p. 18.

3 Bigge 1966, p. 92.

4 Shaw 1967.

5 For a more detailed history of drought policy in Australia, see Botterill 2003b.

6 Grogan 1968, p. 309; Gerritsen 1987.

7 Snedden 1971, p. 57.

8 DPRTF 1990.

9 Senate Standing Committee on Rural and Regional Affairs 1992.

10 Etzioni 1967.

11 ACANZ 1992, p. 13.

12 Senate Rural and Regional Affairs and Transport References Committee 1995, p. 58.

13 ARMCANZ 1999a, p. 63.

14 ARMCANZ 2001, p. 33.

15 Truss 2002a.

16 Braybrooke and Lindblom 1963, p. 73.

17 May 1992, p. 351.

18 May, p. 332.

19 Cohen *et al.* 2002.

20 Kingdon 1995.

21 Botterill and Chapman 2002.

22 Aitkin 1985.

23 Schapper 1970, p. 91.

24 ARMCANZ 1999b, p. 4.

25 Amery 2002; Truss 2002b.

26 Lindblom 1965.

4

Media representations and public perceptions of drought

Åsa Wahlquist

Most city Australians have little knowledge of life in rural and regional Australia. Most of what they do know they learn through the media. However, communicating the reality of drought presents a great challenge for the media. Most are city-based, with little conception of the complexities of the experience of drought.

For example, radio and television weather presenters largely define good weather as the absence of rain. An increase in prices, due to drought shortages, hardly affects most city dwellers, with food and fibre a comparatively small part of the average budget. It is only when water restrictions kick in, as they did in Sydney in 1995, and again in 2002–03, that city people begin to take cognisance of long dry periods. Most city people have little under-standing of modern agricultural production, or of the Australian climate and the significant role that El Niño plays.

A good parallel is city Australians' reaction to bushfires. Though all the state capital cities have large adjacent areas of bushland, the outbreak of bushfires—the worst of which are linked to El Niños, such as Ash Wednesday in 1983, the Sydney bushfires of 1994 and 2002, and the Canberra bushfires of 2003—routinely provokes first astonishment, then calls for the area to be made bushfire-proof. This is despite the fact that Australia is a continent with a long fire history, and a vegetation in many areas that is superbly evolved to live with fire.[1]

The reality is that Australia can be neither bushfire-proofed, nor drought-proofed. Learning to be Australian means learning to live with El Niño and its droughts and bushfires.

Just over one century ago, when there were fewer than four million Australians, more than half of them lived in the bush.[2] In the years leading up to World War II, most urban Australians had a relative on the land. Life on the land was deeply familiar to most city dwellers; they knew when wool prices were high and wheat prices low; when there were droughts and when there was rain.

The great waves of migration began with post-World War II European refugees and continued from Greece and Italy, from south-east Asia and the Middle East. By 1996, over 16 per cent of Australia's population had been born overseas. While many might have come from rural backgrounds, they made their homes overwhelmingly in the metropolises of Sydney and Melbourne, and to a lesser extent the other capital cities. By 2001, 64 per cent of Australians were living in the capital cities, making Australia one of the most urbanised countries in the world. About 86 per cent of Australians live in the cities, or within 80 kilometres of the coast. That means that these days just 14 per cent of Australians live in inland Australia—in the bush.

Tim Fischer represented the rural electorate of Farrer, that lies along the NSW side of the Murray, for 17 years. The former Trade Minister and Deputy Prime Minister believes city people do not understand drought.

Albeit now or then, they do not, and the disconnect grows between capital city Australia and country Australia, the links when almost everyone had a country cousin no longer exist physically, with the generations becoming very urbanised. It really is becoming a very difficult situation because the passage of many generations have no direct country contacts whatsoever, and that is also adding to the burden.[3]

So how do urban Australians, or those who live along the great coastal arc from Barwon Heads to Cairns, who do not have the country cousins of several generations ago, learn about life in country Australia? The answer is through film and books, but mostly through the media. But this presents some problems. Films, for example, set in rural Australia tend to emphasise bizarre characters and the stark geography. From *Crocodile Dundee* to *Priscilla Queen of the Desert*, from *Picnic at Hanging Rock* to *Rabbit Proof Fence*, the bush signifies alienation, struggle, even terror and a gulf of difference.

Australian literature is primarily urban. The most successful novels that are country-based, like Murray Bail's *Eucalyptus* or Michael Meehan's *The Salt of Broken Tears*, or Peter Carey's *True History of the Kelly Gang*, are set at least one generation ago. Television drama, when it is set in rural Australia, tends to continue the theme of rural Australia stuck in the 1950s. Generally, farmers are portrayed as poorly educated men, who wear big hats and speak slowly, and who are struggling with the modern world: people engaged in a lifestyle, not a business. There is little evidence of the reality that one-third of farmers are women, most are computer literate, around 40 per cent belong to Landcare, and thus are more active conservationists than many city people who like to think they are green, and that broadacre farmers made a very respectable annual productivity gain of 2.5 per cent over the 20 years to 1996-97.[4] In fact, grain growers made an astonishing annual gain of 3.6 per cent, a figure most Australian industries can only stand back and envy. Australian farmers are among the most efficient—and least protected—in the world, but nowhere in popular culture is this portrayed.

Most city people gain their understanding of rural Australia from the media: newspapers, television and radio. But most members of the media reflect their country: they are over-whelmingly city people, with little understanding of country life. The problem is exacerbated by the lack of specialist rural reporters in most metropolitan media. From his family's fine wool and beef cattle property in Glen Innes in northern NSW, Tim Hughes both observes the media, and contributes to it, as a farmer and freelance journalist. In an article published in August 1999, Hughes argued:

Regional—not just rural—Australia appears to be getting less
exposure in the metropolitan media; what does run is
increasingly negative and covers a narrower subject base. In
many metro dailies, and even the nationals, the number of stories

from some overseas countries is often greater than the combined news of all non-capital-city Australia.[5]

Hughes quoted chief analyst of Media Monitors Analytical and Research Services, Milton Hill, saying that, although there was no data, he believed there was much less rural news being carried than 20 years ago. He said the bush only got coverage 'if it's drought, flood, or lately Wik'. Hill suggested this in large part mirrored demographic movements, though he also pointed to 'agenda clusters', 'whereby editors at the majors are increasingly covering the same stories as each other, but fewer issues overall'. More recently, Hughes argued that it takes a really hard sell, particularly if the story 'doesn't have a novelty hook to it', to get a chief of staff or a news editor interested in a rural or regional story: 'Perhaps the political changes that have come about with the rise of rural independents and One Nation and the debates about the privatisation of Telstra, have helped to put regional Australia back on the agenda a bit'.[6] And he laments the tightening financial imperatives of news outlets, which has virtually seen the end of the old system of country correspondents or stringers, and reduced travel. 'If (television) crews fly out, they do so quickly and on an *ad hoc* basis, there is no time to develop a relationship with the subject'. As a result, Hughes says 'there is a trend to stereotype rather than present the diversity of the country. I think that newspapers are increasingly focussed on lifestyle stories and the realities of regional Australia just can't compete with a readership that seems increasingly obsessed with the cult of celebrity.'

Hughes quotes rural journalist, Anthony Hoy, who has worked for the *Sydney Morning Herald*, and more recently *The Bulletin*, who said:

> *One of the major frustrations for a specialist rural writer on a metro paper is that most senior editorial staff, and sub-editors, have an urban or international background, with little or no understanding of their significant, neglected constituency. For most, a visit beyond the metropolitan limits, apart from a skiing holiday or a trip to the wineries, is the exception rather than the rule.*[7]

My own experience, as the rural writer for the *Sydney Morning Herald* from 1991 to 1994 was that the operative word in the title was 'Sydney'. However, working for *The Australian*, a national newspaper, has been a very different experience, with *The Australian* seeing reportage of rural Australia as a fundamental part of its news gathering. As I argued to a forum on the reporting of rural Australia in October 2000, a clear distinction should be made between the metropolitan and national media.

Generally speaking, the national media at least acknowledges rural and regional Australia. They understand it is integral not just to the national economy, but also to who we are. But read most metropolitan papers, and you would never know that one-third of the population lives outside the capital cities. There is next-to-no-coverage of country social issues, of regional problems, or even the highly productive agriculture and mining industries, which despite being unfashionably old economy, still underpin the national economy. In fact, were the metro media to look closely at those industries they would discover a sophistication that would shock the cotton socks off many city folk.[8]

I cited as evidence the importance of rural-based issues such as Pauline Hanson's One Nation party, native title, land clearing and salinity, and genetically modified food.

Then there is the growing political importance of country Australia as regional voters learn to flex their muscles. Most of these issues have been poorly handled by the city media. In some cases they did not know what was happening, or they underestimated the forces at work. In other cases they simply failed to understand them. Perhaps the biggest surprise for the city media was the defeat of the Kennett government in Victoria last year [1999]. Yet the signs were there for those with a eye to reading them.[9]

Radio is a good medium for relaying information: but there is a vast difference between the high-quality information coming from ABC Rural, with its many specialist reporters, and the shock jocks, who though often sympathetic (like both Alan Jones and John Laws, in Sydney) do not have the depth of understanding of the specialist reporters. An outstanding example of this was Alan Jones' call, in late 2002 for some of the large coastal rivers, such as the Clarence in northern NSW and the Burdekin in Queensland, to be turned inland. It was a call made with (misplaced) sympathy for drought-stricken country people. It was roundly dismissed as impractical, at the least, by a large group of people from the scientists of the Wentworth Group, to Prime Minister John Howard.[10]

Bob Collins was the Federal Primary Industries Minister from 1993 to 1996. As a Senator representing the Northern Territory, he was acutely attuned to media coverage of the regions. Collins said there is a strong view that only the ABC 'gives the bush a fair go'.

> *A lot of rural people have a very, very fond spot for the ABC, programs like the* Country Hour *for example, which are required listening for everybody out there. They have this fond appreciation of the job the ABC does for the bush. There is a very strong view that it is the only bit of the media that gives them a fair go. There is a real concern about the town and the bush in terms of media. The overall impression as to how they saw the media was not particularly country friendly in terms of mainstream media, with the exception of the ABC.*[11]

Neil Inall is a veteran rural journalist. He was also chairman of RASAC, the Rural Adjustment Scheme Advisory Council, which made recommendations on drought relief to the Federal Minister for six years from 1993. Inall says the general media's approach to reporting on rural Australia is 'totally opportunistic'. He cites its approach to drought.

> *It sees a few heart-wrenching stories that it can run, that gives them good pictures, so that fills up the paper for a few days and then they forget it. When there is a good drought-breaking rain they will go out and talk to Joe Bloggs, who is having problems now with excessive water and the sheep that got washed away in the creek. It is very opportunistic.*
>
> *The media really has done a very poor job in analysing the occurrence of drought in Australia, the science behind it, the whole matter of the El Niños, the Southern Oscillation Index, the ocean temperatures, following up on CSIRO research on drought. I don't think the media has done a good job about this at all.*[12]

Inall concedes that urban editors are more likely to be interested in human distress stories than the science behind drought. But he is particularly critical of the major rural media in Australia.

The Rural Press and the ABC really do little or nothing of the sort of thing I am talking about: the science, the policy options. There are no great discussion programs or articles analysing the policy, analysing the science, analysing how Landcare is very much part of drought management, drought preparation.

The major rural media, the print media and the broadcast media have not served agriculture, have not served Australia well, because they have ignored the science, the analysis of policy, the inconsistencies by governments, both Federal and State in this whole area.

One of the big problems in covering rural Australia, and particularly drought, is the whingeing cockie syndrome. Bob Collins said farm leaders are very, very sensitive about it.

There is a view around that primary producers are always bitching about something, there is always a situation where there is either too much rain or not enough rain, or commodity prices are no good, or there is not enough help for the bush. I found as Minister that there was a real sensitivity about that, with the peak industry leaders, that they were anxious to ensure that they tried to not perpetuate that.[13]

This is a perception backed up by New South Wales Farmers' Association research in 1999, into attitudes to rural Australia. The study found that, while 30 per cent of the respondents never visited rural areas, they still chose the farming sector as being the most important industry to Australia. But more than half felt that 'the only thing you hear about farmers is problems'. Even more farmers agreed with that statement.[14] It is an interesting perception from a group that wields a very strong public relations arm, that leads the call for drought support for farmers. Hammering this attitude is a favourite theme of Tim Fischer's. In January 1999, after a four-state two-territory outreach tour, where he saw many farmers bouncing back economically, he called on farmers to drop their negative talk: 'The ability of many Australian farmers and some, not all, farm leaders to maximise the negative mantra has reached the stage where it is counter-productive and damaging for the future of the agricultural sector of the Australian economy'.[15]

The reason farmers complain, according to Fischer, is because 'it's part of exercising leverage on State and Federal governments'. And the story of drought assistance, the media coverage and government response certainly fits that bill.

The only way most city people learn about country people and how they manage—or fail to manage—drought is through the media. This presents huge problems. Drought is very complex, but television is not well suited to intricate arguments. Television current affairs thrives on individuals laying blame: the person who blames the government for not helping them during drought is infinitely more interesting to the TV program than someone admitting they had not prepared adequately. And then there is what Tim Hughes calls the iconography of drought.

> *It is still the same images that get thrown up the whole time: the dusty drover on the stock route, or the parched earth with the ram's skull piercing it. Those sort of images, there is almost a sense of romance in a way. It harks back to a grand artistic tradition, people can make a psychological link with Tom Roberts or Nolan or Drysdale. The images you are never going to see in the mainstream press, are going to be of sheep having to be shot or a drafting race of poor cattle on the way to getting the lead bullet at the meatworks because they are more confronting.[16]*

This, argues Hughes, makes reporting the big debates about restructuring farm business systems and taxation, the very concept of drought, more difficult. Instead, Hughes says editors prefer the language of battle.

> *We like a battle, in national culture it is character building. This battling the drought, we battle it rather than accept it, and continue to battle it. It is not how do we live with it, or work with it?' But OK how do we arm ourselves against it, whether it be the form of drought subsidies or anything else? Is it like Gallipolli where it is inevitable, we are always going to be the losers, but the battle is part of our national character and the loss, perhaps, too?*

Hughes admits that conflict is the stuff of daily journalism, 'because conflict is what creates the drama which makes it interesting for readers'.

Former member of the Queensland Drought Secretariat, Dan Daly, has written a biting analysis of Queensland's drought policy, 'Wet as a Shag, Dry as a Bone'. He argues that in a climate of farmers asking for drought handouts: 'Media reports of extreme situations on properties distorted perceptions about the severity of the drought and softened attitudes for an early declaration'.[17] By the time of the 1994–95 drought, the public image of farmers was that of the beleaguered battler. Over the past decade, the Australian farm sector had suffered a series of blows. The high interest rates of the 1980s had resulted in a series of high-profile television reports of banks forcing farmers from the land. Some farmers were caught with foreign currency loans that had spiralled into impossible debts.

Then in 1991, wheat prices fell to an all-time low, due to the depressing effects of the subsidy war between the United States and the European Community. The wool floor price ended, under the weight of the wool stockpile that took a decade to sell. In the space of four years, wool—which was at that time the most widely grown commodity in Australia—fell from 1269 cents per kilo clean in 1988 to hit 396 cents per kilo in 1992.[18] By April 1991 the first NSW drought stories were appearing. Older farmers reported that not even the depression was as difficult as the times they were facing.[19] Rural land values plummeted, trapping many on the land.[20] By early 1992, the Australian Bureau of Agricultural and Resource Economics was forecasting that the financial year would be the worst for 40 years. Bureau Executive Director, Dr Brian Fisher, said the downturn reflected weak commodity prices and the drought, with the net value of farm production forecast to drop to a record low in real terms in 1991–92.[21]

The joke in country circles was: 'What is the definition of rural child abuse? Leaving the kids the farm'. By this stage, farmers were typically portrayed by the media as long-suffering, set upon by fate, and often bewildered by the changes that had occurred. And the big drought of 1994–95 had not yet begun. In 1992 there was a dramatic change in drought policy. As discussed in Botterill's chapter in this volume, in 1989, drought was taken off the natural disasters list. In mid-1992, as drought was affecting much of NSW and parts of Queensland, then Primary Industries Minister, Simon Crean, argued the case for a risk management approach to drought. Collins, then Senator Collins, took over the Primary Industries portfolio from Crean in December 1993. He faced three immediate problems.

> *I hate to say this, but it is true. We came in at a time where we had had a whole run of very large interest rate levels, they were by today's terms extraordinarily high. That was absolutely crippling primary producers in terms of the loans they were needing to get to pay for stock feed, and stuff. It was the absolute triple whammy of high interest payments, low commodity prices and devastating drought, leading to a cash-poor farming community, where people were getting to the point where they were walking off their properties.[22]*

In other words, the National Drought Policy, which demanded farmers prepare for drought, came into effect in the worst financial period for farmers in 40 years, when few had had the ability to put any resources aside. Farmers were already suffering a cash drought. As 1994 ticked by, the area affected by drought grew. Collins said:

> *We were very conscious of the fact that we had a major national disaster on our hands, and that particular drought was the worst ever. It was a national drought. It devastated the whole of Queensland, most of New South Wales and Victoria, and certainly South Australia, and at one point we were even paying drought assistance to areas in Tasmania.*

The media, Collins said, was critical in communicating the depth of the drought.

> *It was absolutely critical, there was no doubt about it. The key I think always to getting the right story across is to get people out on the ground. There is no question about that. Writing these stories from your desk in Sydney or Melbourne just isn't the way to do it. If you, as we did, take key media people to places like the Darling Downs, which traditionally is the crème de la crème of Australian agricultural country, and they actually see areas of the Darling Downs that at that point in time had just turned in their fourth successive crop loss you get the message home.*

Collins singles out the Farmhand fund, sponsored by Channel Nine's *A Current Affair* and spearheaded by Ray Martin. 'You could hardly have given a more prominent or sympathetic high media profile to the devastation that was hit by that drought, that Farmhand gave'. Ray Martin moved to *A Current Affair* in 1994, after a decade hosting *Ray Martin At Midday*. On *Midday*, Martin said they covered a lot of bush stories, '*Midday* was almost like the bush telegraph.'

> *I am one of those who believes that even though most of us live close to the city, it is part of our ethos. And going to try and get money for the Farmhand appeal, I felt confident that the city really believes it is part of the bush even though it rarely goes over the mountains.*[23]

Reporter Paul Lockyer brought back to *A Current Affair* some harrowing accounts of the drought, 'and it struck a chord,' Martin said. 'He and I thought there was something bigger there'.[24] The idea of Farmhand came to Martin, not as a bolt of lightning, rather 'it just seemed to be so obvious we wondered why it had taken so long. I thought a) it was a good project, and b) it was a very promotable thing'. Martin also thought it would give *A Current Affair* a heart.

> *Lots of current affairs programs, I think, have an image of being arrogant and not understanding its market. The image of* A Current Affair *when I came back to it, was that we tended to chase people down streets, and we tended to put our foot in doorways. I don't think that was true, but that was the image. I think we needed to be a little more viewer-friendly. And there was the fact that farmers were desperately in need of a helping hand and a bit of attention.*

Backed by the good ratings the drought stories received, the *A Current Affair* team spoke to the National Farmers' Federation, which contributed a staff member to aid distribution. The National Bank came to the party to collect the money, while the Salvation Army and The Smith Family distributed it. Farmhand ended up collecting $19 million. Martin said:

> *What surprised me was the way it took off. I don't think we really have a philanthropic tradition in Australia, we volunteer*

*and go and help in bushfires, and the surf lifesaving, but
generally we are slow to put our hands in our pockets. But this
was one where schools were sending us $1000 they had raised
at a fete, and golf clubs were saying, 'we had a whip around
and here's $431'.*

Martin said that bipartisan support from politicians helped. And
he believes *A Current Affair* and Farmhand influenced the delivery of
drought policy.

*I have no doubt on that. This was a campaign. We had 2UE
and other radio stations, they weren't part of it, but they picked
up and supported it. The media combined, the (Daily)
Telegraph, for example, and News Limited picked up and
supported us as well. And like any good publicity campaign, the
pollies realised there was votes in it. They were on a winner and
they didn't have to appear to be ruthless bastards. Even the NFF
was saying they couldn't quite believe the Labor government was
so onside, and wanted to cooperate.*

Martin said Farmhand was an example 'of how the media can
almost be Australia Inc., and we can pull together on those things.
It was a lay down misère, you couldn't fail as long as it was done
with some sort of honour. It was a win/win for everybody'.[25] In
October 2002, the exercise was repeated, though perhaps with less
passion, raising $20 million.

In one three-week period, in October 1994, Farmhand gave out
more than $3 million to 1500 farmers. Farmhand national coor-
dinator, Doug Miell, estimated it had helped between 12 000 and
14 000 people, with gifts ranging from pencils donated by kinder-
garteners to $100 000 cheques from corporate Australia. In the
deeply affected shire of Bourke in western NSW, which had been
drought affected for four years, about 90 per cent of the shire
received some assistance from Farmhand. Rural Counsellor, John de
Bomford said 'A lot of people felt they had really been left to their
own demise and that the country didn't realise what was
happening out here. It gave them a real boost'.[26]

While Farmhand built up momentum, it highlighted the
problem Bob Collins was facing. The new drought policy was
falling short. The problem, Collins said, was that farmers 'were

technically asset-rich, but totally cash poor. They had no cash income at all coming into the house because they didn't qualify for any social security payments'. Collins went on television, displaying the parcels that were being sent out to farmers, 'with basic necessities like soap and toothpaste'. Many farmers were so cash poor they could not afford such essentials, 'and Farmhand got that message across'. Prime Minister Paul Keating was seen to be most unsympathetic, after making the statement that drought was just a normal part of the Australian landscape. To be fair to Keating, this was simply restating National Drought Policy. But the drought of 1994–95 was looking exceptional.

Collins was faced with the fine line between propping up improvident farmers and distributing political largesse, and ensuring social equity.

> You can imagine how sharp edged those concerns were in a Labor cabinet, we had quite a lot of robust debate, because there is a strong perception of 'why should people that are sitting on an asset worth millions of dollars, technically, with a homestead and retirement properties on the coast, get access to public money?' It's a tough argument.

It was an argument that was swayed by the media portrayals of suffering farmers.

> Correct, and the reality of the fact, you would go to cattle properties in western Queensland, where you put quotes around asset rich—at some point in time, sure the asset is worth something, but at the time the drought hits, it is worth nothing, the thing is not saleable. And they have no money at all. It is unquestionably in Australia's interest that the skills and the preparedness of people to actually live in those places and to turn off productive results from that land, it is in the national interest to tide those people over. I thought there was an overwhelming case to tide people over because it wasn't in our interest to have people walk off the land. And I ran those arguments in cabinet.[27]

Collins asked the Prime Minister to go bush, and see the devastation for himself.

It was a comprehensive trip, where the Prime Minister went to the worst affected areas of Australia and saw on the ground the extent of the devastation. It was a real eye-opener for him, and he appreciated the fact that this drought was well and truly beyond the normal drought.

That trip was important for two reasons. One was to actually involve the Prime Minister of the day directly in understanding the extent of the devastation of the drought. I knew as a Labor minister in a Labor cabinet wanting to go in when times were tight, and ask for a huge amount of money for a constituency which had never voted Labor in its life and never would, was going to be a hard ask. I knew we would need Keating and the only way we could get Keating was to take him out and show him the drought. That was important.

Keating came to the party and I would never had got that drought package through cabinet if it had not been for the PM's support in cabinet. But it was also critically important from a media perspective, because if you take the Prime Minister on a high-profile drought trip, you are guaranteed of prime-time coverage, which is what we got.

The result was Keating's stirring woolshed statement on September 1994, broadcast on prime-time television, when he extended his election promise not to leave the unemployed behind, to the country people of the nation. 'He was brilliant,' Collins admitted, 'it was Keating at his very best'. Along with the statement was a new, generous welfare package. Collins says: 'It was widely praised, I was amazed at the response. There were very few people who bagged it at all. It was publicly credited with saving a lot of people from actually walking out the gate, and it was delivered in an extraordinarily short period of time'.

According to Collins, the media played a centrally important role.

It has to be acknowledged that there is a widespread view of the whingeing farmer, that farmers are always complaining about something. It is those graphic images that were able to turn around that perception where people say, 'OK we acknowledge the fact that this is something special, this is not just a normal problem' and it is the media that does that. You can't do it without the media.[28]

Neil Inall said Farmhand was an extraordinary story in itself.

I think it helped some people who were in real need; in many cases, of course, people who had done no preparation whatsoever to help themselves. Ray highlighted the problem, but, again I think it was about that culture of 'come on help me, I am a farmer, and because I am a farmer having a dry time I need to be helped'. I think we have this dependency culture when it comes to the climate and Australian agriculture. I am not saying that there aren't times when governments and the community ought to help out, but because drought is such a part and parcel of our life, we ought to be doing everything possible to encourage people to prepare for it.[29]

Inall was critical of the paucity of stories about farmers 'who prepare for drought and who get through more or less unscathed. And that is a hell of a pity, because apart from anything else, when there is assistance given, those people don't get any, it is the people who do bugger-all to prepare who have always got the assistance'. As the rural writer for the *Sydney Morning Herald* during the 1994–95 drought, I would regularly be rung by good farmers who were angry that their improvident neighbours were receiving government aid. But when I asked the caller if I could write them up, to tell the story of a farmer managing capably through the drought, the answer was almost invariably no. They did not want to be seen to be critical of their neighbours.

Then I got a call from George Gundry, of Willaroo, near Goulburn. Gundry rang not to criticise his neighbours, but because he wanted to publicise his new management system. The article, resplendent with a photograph of George and his wife Erica, enjoying a picnic on a well-grassed paddock, ran on 7 December 1994. It reported that the Gundrys had actually been making hay, while their neighbours were drought declared:

If I can make hay in drought, when I haven't made hay for years, is it a drought?' he asked the Herald. *In NSW, drought is determined on the basis of available pasture. Willeroo has received the same rainfall as the neighbours: the difference is the Gundrys' pasture management and decision-making. The Gundrys practise time control grazing, a system based on husbanding their most precious resource, their pasture. By studying his paddocks, he noticed the first signs of drought months before the area was officially drought declared. 'I recognised the signs of deteriorating season in the autumn,' he said. 'We heavily de-stocked at that stage, 2000 sheep out of 10 000. It was a big decision.*

I have a set of rules I follow and one of them is 'I shall not substitute feed grazing animals in the drought.' Handfeeding is economically very risky. You don't know when the drought is going to end; in other words how long you are going to feed for.

The other problem Mr Gundry sees with handfeeding is 'you usually reach the stage when the ground is bare anyway'. It is critical to his management that the property never loses its cover of pasture.[30]

The response to the story defied conventional wisdom: it was not only widely read, it even received coverage on radio 2BL in the morning conversation between the late Andrew Olle and Paul Lyneham. It showed there was an interest in the experience of drought beyond the suffering farmer. On 31 December 1994, page one of the *Sydney Morning Herald* 'screamed': 'drought now the worst in history'. It reported that an unprecedented 98 per cent of NSW was drought declared, and that the national wheat crop was halved.[31]

Hughes argues that the system of drought declarations (used only by NSW and Queensland; the other states rely totally on the Federal scheme) gives drought reporting natural disaster overtones.

The way it is reported by the media that, all of a sudden, overnight, when an area is suddenly drought declared ... It is not as if the warning signs are ever really given. It is almost as if it happened overnight, so perhaps there is a similarity with the reporting of the natural disasters that happen overnight.

He also argues it does not help the public to understand that drought is officially no longer a natural disaster.

That declaration means that, the day an area is drought declared may change the whole eligibility or perception of farmers and government agencies. But the reality is it doesn't change overnight, it has been a long process and will continue to be a long process for many years after the rains have fallen. This whole drought declaration thing does give news decision-makers and the general public the view that it is switching it on and off like a light switch. It is not like that in reality.[32]

In January 1995 good rains fell on much of NSW. On 23 January 1995 the lead photograph on the *Sydney Morning Herald* was of a gum-booted farmer surveying the swelling waters of the Darling River.[33] The photograph was actually taken on the same spot where Premier Fahey had posed on what was then the dry riverbed, five months earlier.

Overall, Collins said the media:

... were largely supportive and supportive of the fact that government assistance was necessary. People weren't bagging it, the mainstream media generally was very, very supportive of all that, and were happy to give prominence to the drought package when it was released.

We got a lot of assistance from the media, and I thought the coverage was intelligent, informed in the main, balanced and pretty supportive. And it was critically important to both the formulation and the delivery of the drought policy that was eventually delivered.

It was also critically important in publicising the features of that policy and how it could be accessed, and there was very large take up of that policy, so I thought they played a very, very positive role.[34]

As the rivers flowed again, and pastures grew, there was some analysis of what had occurred. NSW Agriculture has produced a poster called 'Drought: Advance and Retreat'. It displays 20 years of drought declaration maps. Not one year on the poster, which begins in 1972, has entirely escaped drought. Only three years—1974, 1984 and 1989—qualify as almost drought free. Some years, in the early 1980s and 1990s, are black with droughted areas. The lesson is: drought is endemic in this state.[35]

ABARE estimated the 1994–95 drought reduced the net value of farm production by $1.95 billion with a flow-on effect to the rest of the economy of $3.3 billion, equivalent to knocking 0.75 percentage points off economic growth that year.[36] Overall, the federal government committed $590 million, assisting 10 400 farm families, the Queensland government committed $90 million, while the NSW figure was $202 million. Charities provided over $20 million, and at the end of 1996 they were still handing out 'something like $35 000 a day', according to NSW Agriculture drought relief coordinator, Geoff File. With the drought effectively over, questions were starting to be asked. According to File, there was contention over the drawing of the assistance lines, and who did or did not become self-reliant. 'A lot have become self-reliant and they were the ones who didn't get the assistance', File says. 'There is a realisation from the farming organisations that farmers have to move to self-reliance.'[37]

Tim Fischer says that, as long as the hand of government can be extended towards them, there will be political incentives for farmers to complain, and pressure on those who are managing to remain silent. But, he says, there are changes afoot.

I notice some of the generational change farmers are much more media savvy, and much more angry if their colleagues cannot handle a period of twelve months drought and haven't calculated that into their farm budgets and reserves. Beyond twelve months it is another area and you get into the legitimate area of exceptional circumstance and assistance.

This is Australia. This is the second driest continent in the world, and this cowboy farm management approach of push the limits at all times and scream the moment there is a period of dry, or even a drought of a one-year cycle, ought to be exposed for what it is: very damaging to the environment and a less-than-smart business approach.[38]

Neil Inall suspects a lot of urban people have become 'very cynical about droughts: "Oh god, farmers whingeing again and why in the hell should we be helping them if they don't prepare for it?"

The system has perpetuated that by governments giving assistance when it gets a bit dry.'[39] He is frustrated with the lack of communication of the 1992 decision to reduce assistance, and that farmers should look after their own risk management. 'It hasn't got out at all, hardly at all'. It is vital, Inall says, to get good information about drought, and drought policy, across to the general audience 'because they are taxpayers and sooner or later there is going to be a government which says "we are not going to help at all"'.

It is a terrible irony that in this country, with the most variable rainfall in the world, the language of war and disaster, and the imagery of suffering dominates drought coverage in the media. This portrayal has clearly had a powerful influence on governments and policy. Drought is an inherent part of life in Australia. Until the media understands that fact, and begins to celebrate the survivors, those who have learnt to manage drought, rather than those locked in a failing battle with it, it will be difficult to conduct a fully informed, rational debate about drought policy in this country.

1 Pyne 1991.
2 These population figures are sourced from various Australian Bureau of Statistics publications, including Hugo 2001.
3 Tim Fischer, interview July 27, 2002
4 Martin *et al.* 2000, p. 51.
5 Hughes 1999.
6 Tim Hughes, interview 26 July 2002.
7 Hughes 1999.
8 Wahlquist 2000.
9 Wahlquist 2000.
10 Megalogenis and Wahlquist 2002.
11 Bob Collins, interview, 23 July 2002.
12 Neil Inall, interview, 27 July 2002.
13 Bob Collins, interview, 23 July 2002.
14 Hughes 1999.
15 Wahlquist 1999.
16 Tim Hughes, interview 26 July 2002.
17 Daly 1994, p. 94.
18 Wahlquist 1995.
19 Wahlquist 1991a.
20 Wahlquist 1995.
21 ABARE, Outlook 1991, Brian Fisher's address.

22 Bob Collins, interview, 23 July 2002.
23 Ray Martin, interview, 26 July 2002.
24 Ray Martin, interview, 26 July 2002.
25 Ray Martin, interview, 26 July 2002.
26 Wahlquist 1994.
27 Bob Collins, interview, 23 July 2002.
28 Bob Collins, interview, 23 July 2002.
29 Neil Inall, interview, 27 July 2002.
30 Wahlquist 1994b.
31 Wahlquist and Kidman 1994.
32 Tim Hughes, interview, 26 July 2002.
33 Wahlquist 1994b.
34 Bob Collins, interview, 23 July 2002.
35 Wahlquist 1995.
36 ABARE 1991
37 Wahlquist 1996.
38 Tim Fischer, interview, 27 July 2002.
39 Neil Inall, interview, 27 July 2002.

Australian drought as lived experience: Social and community impacts

Daniela Stehlik*

Despite most Australians agreeing that they are only too aware of what drought is, surprisingly little Australian research has been undertaken to explore the experience of living through a drought with those most affected—farm families.[1] As other chapters have detailed, drought is an integral part of the Australian landscape, and most parts of the continent at one time or another are drought declared. Drought is current news. For example, as I write this (July 2002), south-east Queensland is reporting its driest season for 60 years, and *The Weekend Bulletin* headline on 20 July reads: 'Drought tightens grip' and identifies 'nine shires in Central Queensland that have been drought declared'.[2]

The drought of the 1990s, following closely on the experiences of the 1980s, and touching many areas, provided a unique opportunity to undertake research *as it was happening*. Combining as it

---

* **Acknowledgements**: I thank Dr Frank Lewins for his insightful and helpful comments in review and for his kind permission to incorporate them in the final version of this chapter. I also thank our editors and the other members of our 'chapters collective' for their reflective and helpful suggestions.

did with falling commodity prices, high interest rates, and withdrawal of services (not just human services, but also private-sector services such as banks), this experience undeniably changed many traditions in rural Australia. As a way of enabling a more evidence-based policy approach to future drought management, in this chapter I will take a people-centred focus by asking two questions. First, what is the 'essence' of living through a drought, for individuals, families and communities? Second, why is acknowledgment of lived experience important for scholars studying drought and how can these lived experiences be drawn on by the policy-makers of the future?

Lived, or self-experience challenges a universalistic approach that tends to homogenise. Public policy, designed for the many, under great stress and in times of turbulence, often assumes a homogeneity where none exists. It tends not to challenge 'taken for granted' assumptions. Yet policy that recognises difference and actively works to incorporate heterogeneity is a challenge to create and to deliver. This chapter suggests that how we as researchers, policy-makers, students and professionals can come to understand drought through a lived experience framework means that we need to be reflective, insightful and sensitive to language, rather than rely on a system or a theory.[3] There is a place for self experience, as there is for empirical reality and for theory. Greig, Lewins and White[4] suggest the following schema which identifies theory, self experience and empirical reality as three sides of a triangle, an approch which enables self experience to take its rightful place

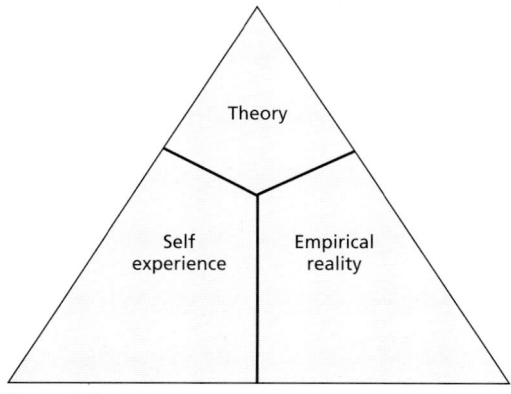

Figure 5.1

alongside that of theory and empirical reality. We took this approach in our study.

By drawing on what has been described as a *social construction of drought*, in this chapter I will use interview material as evidence to enable an understanding of the concept of *resiliency*—at the individual, partnership, family and community levels. This concept of resiliency lies at the heart of social cohesion that in turns enables the communal and individual strength to meet the challenges drought presents.

The chapter is structured first, to identify what is meant by resiliency, and second, to briefly explore the nature of evidence-based policy and the importance of primary sources in developing such evidence. Then a brief description of the research undertaken to develop this analysis is provided. The major section of the chapter draws on this evidence to detail the way in which resiliency can act to meet the challenge of drought and how such resiliency then enables social cohesion. Finally, the chapter makes some broad suggestions as to how policy-makers can themselves enable such cohesion by drawing on experiences in developing future policy for rural Australia.

Living the experience

It appears almost commonsense to say that drought becomes a crisis only when it affects communities and families, yet the drought of the 1990s challenged some deeply held assumptions about the nature of drought and its impact on the rural sector. One such assumption was that exceptional droughts eventuated every 25 years or so,[5] yet in Queensland, many farm families were only just recovering from the drought of the early 1980s,[6] when their districts again became drought declared. Another assumption was that overall drought policy was designed to respond to 'normal' droughts and this required a 'self-reliance' approach from rural people.[7] In this way, such policies attempted to 'strike a balance' rather than provide a 'one-size-fits-all' response. However, such policies also tended to assume the length of the drought period as relatively short—no-one was prepared for the many years of ongoing drought as was experienced in the 1990s—as many as six seasons in some areas.[8] It was also assumed that drought was a relatively slow phenomenon and that [it] could be anticipated and thus people could prepare for it.[9] Another deeply held assumption was that 'community' is drawn together in some way during times of crisis. Indeed, many policies are themselves predicated on this assumption, one which relies on perhaps utopian ideals of community; that is, of everyone pulling together, putting aside their differences, and working towards a common goal.[10] This ideal could be seen in much media reporting at that time. Again, the drought of the 1990s challenged this as its sheer length and breadth acted to create a fragility around such potential cohesion. It challenged the claim that people behave in a uniform manner as if there was some 'collective mentality' that existed simply because people farmed.

The challenges to these assumptions emerged from experiences discussed in data collected during research conducted in Queensland and New South Wales during 1996–1998 with farm families.[11] This project, funded by the Rural Industries Research and Development Corporation and the Land and Water Resources Research and Development Corporation, and conducted by researchers at Central Queensland University and Charles Sturt University, established the data from which this chapter is drawn. The project—*Farm Families' Experience of the Drought of the 1990s*— was the first of its kind to be undertaken with those affected while drought was still being experienced, not just relying on hindsight, or previous memory. This made the findings even more immediate, and in some cases very poignant, as people told their stories and reflected on the challenges they were facing. Importantly, at the time of the interviews, no climatologists, policy-makers or farmers were able to predict how much longer the drought would continue.[12] One prediction that was made at this time—which has come to pass—was a 'reduction in both official and public sympathy' for those affected by the drought. Writing in 1988 Heathcote predicted:

> ... I suspect that there will be even less public sympathy for any future victims of drought who had disregarded the warnings, and more official pressure upon resource managers to adopt management strategies which rely less on official relief and which recognise drought for what it is – a constant occupational hazard in Australia.[13]

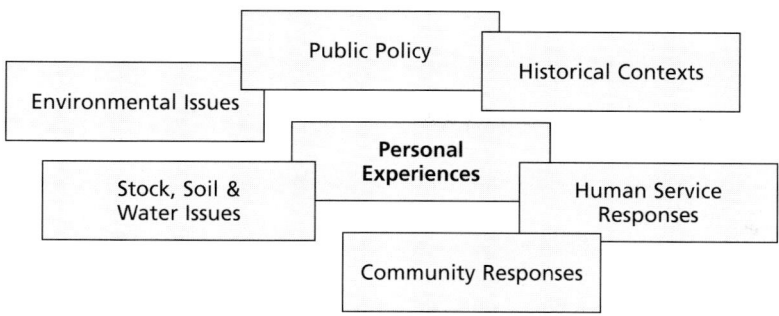

Figure 5.2 Model of social construction of drought Copyright Stehlik, Gray and Lawrence 1999

The research, drawing on the concept of symbolic meanings of drought across farm generations, and how people learn such symbols and how to interpret them, established a model (see Figure 5.2) of a social construction of drought as a 'series of interlinking and connected influences on the personal experiences of individuals'.

In summary, the findings of the overall project were that:

- families are the first line of defence against the hardship of drought
- men and women experience drought differently
- their communities should not be taken for granted by broader society.

The research also found that the majority of producers do

- strive for self-reliance
- manage risk
- try to plan and operate sustainably
- have an environmental consciousness
- undertake 'whole-of-farm' strategic planning.

From a national picture perspective, the research found that the drought of the 1990s in Australia was much more complex an issue than in previous decades, due largely to two important factors (both discussed elsewhere in this book)—first a shift in government policy and second, more sophisticated technologies of measurement. It also enabled an explanation as to why inequality continues to persist. In addition, the research found that:

- drought policy has unintended consequences
- 'experts' can cause additional stress
- human service responses require better integration
- there is a certain cynicism about the media
- rural Australia feels isolated from, and abandoned by, urban Australia.[14]

The study also suggested that while people could not easily plan ahead in such an environment, for many, the interview became a catalyst for much reflection on questions such as: Would they remain on the land? Could they face an experience such as this again? What did it mean for their own health and their children's? How did it affect their plans to provide for the next generation? This capacity of individuals to consider the future, even in the depths of despair, is at the heart of what can be considered as *resiliency*.

Enabling resiliency

Being resilient means that we can 'bounce back' after a shock or trauma. It particularly means that we have in place coping strategies that enable us to emerge successfully from such trauma. International and Australian research suggests that resiliency is formed in families, between individuals, and/or through their own networks, either family or social. Strategies for coping vary across individuals and across families and kinship networks. The research undertaken in the time of drought, found that individuals identified a variety of personal coping strategies; in some cases, relying heavily on their spouse; in other cases, drawing on friendship networks established over many years. For some, professional support became important in enabling coping to continue. For others, planning for the future enabled less anxiety about the present.

Resiliency can also be identified at the community level. Social solidarity and community interaction enables a town or district to maintain its cohesion in the face of hard times. Such resiliency goes beyond what is sometimes referred to as 'communion',[15] or the sharing of communities of interest. It is at the heart of social cohesion and it emerges most powerfully at times of crisis. Findings from international studies identify the important concepts of community support, community empowerment and communal coping as 'broad categories of community protective factors [so that] when a community moves through the process of resiliency it becomes more successful at mastering adversity and change'.[16] As such, resiliency is much more than just social capital, for it acknowledges that there is ambivalence about this cohesion, that it may not be successful in every situation, but that in its very ambivalence it accepts *all* members of the community, even those who perhaps do not have an overt productive contribution.[17]

Enabling resiliency should be the guiding principle behind the policies of self-reliance and sustainability, and all policies developed should ensure that they build such resiliency, not diminish it. However, as the research conducted during the drought showed, many people expressed concerns that they were not able to continue to meet the challenges of the ongoing crisis. In some cases, their anxieties were compounded by a frustration about the formal government definitions of drought, and how these impacted on their capacity to manage the crisis. In other cases, they expressed disappointment in 'the rest of Australia', their fellow citizens living in metropolitan areas, for whom drought may have meant perhaps not watering the garden as regularly, but who, they felt, did not understand the depth of their own circumstances. In this way, Heathcote's prediction had become painfully realised. In addition, some identified the media as contributing to their own capacity to remain resolute or not.

Naturally, government policy also figured in their analyses of coping strategies. The three tiers of Australian government (Federal, State and local), relatively complicated for people to understand in good times, became even more complex as the drought lingered, and policies were developed to deal with it. The human service response to the drought also received some criticism. Regional and rural 'cut backs' and the scaling back of services appeared to occur at the most difficult moments—acting as catalysts for community fragmentation.[18] As a result, social dislocation, feeling disconnected from their existing networks, resulted in a diminished level of resiliency, and for many, the sense of 'doing it alone' became more and more evident as the crisis continued.

Drawing on evidence

Before turning to evidence in more detail, the following provides a brief outline of the project, how the data were collected and its general findings.

In order to capture the differences of production, the research was conducted in two discrete areas: the cattle grazing region around Rockhampton in Central Queensland (CQ) and the wheat/sheep region around Balranald in western New South Wales (NSW). A major assumption underpinning the research was that men and women would experience the drought differently, and thus the project was established to capture that difference. First, focus groups were conducted in both CQ and NSW. These enabled the development of a detailed questionnaire. Then, 103 people (51 men and 52 women) were interviewed on 56 properties in both states. In NSW, 27 women and 27 men were interviewed on their own properties, in their own homes and at times chosen by them. In CQ, 25 women and 24 men were also interviewed in this way. Interviews usually lasted about two hours, with men and women interviewed separately on most occasions. The same questions were asked of both partners. Questions included: the ways in which they perceived and managed drought, financial implications, quality of life and health effects, community support, government policies, and farm and household management strategies.

Once analysis of this interview data had been completed, a subsequent round of interviews was conducted with a smaller group of families in both regions, enabling the development of more detailed case studies. Finally, a report to the RIRDC[19] was prepared in 1998 and published in 1999.

A brief snapshot of those involved in the project provides the breadth of it. Their average age was in the 45–49 year range, with a number in CQ over 55 years. They were experienced farmers with an average of 25 years on their properties, which overall, averaged 9000 hectares—Queensland at 5138 hectares and NSW, at 18 511 hectares. Half of the farms were entirely dedicated to grazing, with only eight farms of the total using less than 50 per cent of their land for that purpose. At the time of interview, 90 per cent reported that their farm was in an area that was drought declared, and while some rain had fallen, only 19 per cent felt that their drought had been broken. The research found 69 per cent of the whole group felt that this drought was more severe than the one in the 1980s. For more than one-third, the drought had either eliminated farm production altogether, or reduced it to its lowest level ever. Importantly, about 50 per cent reported taking the drastic step of selling breeding stock,[20] with around 15 per cent saying that they had taken this option 'extensively'.

There have been wide-ranging publications analysing this project, both nationally and overseas. These have covered topics including gender and drought;[21] drought and community;[22] economic impacts;[23] agricultural restructuring;[24] and social policy and drought.[25] These publications have drawn on the whole body of evidence from all the respondents. Rather than restate these and dwell in detail on the negative factors associated with the experience, in this chapter I use them as context, but draw more particularly on those families interviewed and then re-interviewed—the so-called 'case studies'. This group provides a very detailed examination of the social impact of drought, focusing on individuals, partnerships, families and communities, and emerging coping strategies. By taking up the question of resiliency in each case, this next section will draw on the voices of some of these respondents, and enable them to tell their own stories—although pseudonyms are used in all cases. In this way, this chapter is used to focus on the capacity to respond to the crisis with positive approaches, while at the same time identifying factors associated with government policies and human service responses that, in the opinion of those interviewed, could be better managed in future.

Drawing on evidence to develop policy more effectively has long been understood as fundamental to achieving policies that meet needs. In this research, we understood 'evidence-based policy' to draw on what Barber[26] suggests can be considered as a 'continuum of evidence' from inductive theory at one end of the continuum, to randomised control trials at the other. In other words, that human beings draw inference from a variety of sources. This is related to an important claim that:

> ... evidence is always evidence for something (of a conceptual nature), which in turn is related to the particular vantage point of the researcher. In short, evidence is never 'out there' to be simply gathered; it needs to be interpreted and selected.[27]

The research interviews enabled a richness and depth to our understanding of people's lives to emerge. Barber argues that no one kind of evidence is superior to any other, and statistical 'proof' is not the only data on which policy can be predicated. As the authors of the chapters in this book argue, a balancing of all forms of evidence, including social and community evidence, enables well-grounded policy formulation. In the case of the drought, much evidence of this nature was available to policy-makers; however, as stated above, until our research had been conducted, there had been little attempt to systematically ask producers what their experiences were like, and how they were managing the crisis.

In the following sections, I begin to draw on the experiences we gathered during our interviews. This section includes the voices of those who participated in our research. As such, they are *perceptions* of reality and not an automatic snapshot of their social reality; therefore, the expression 'things that are perceived to be real have real consequences' becomes our guide. In other words, this means the taking up of issues as they are immediately experienced by individuals, rather than as we may categorise them externally, or as we may think that individuals should be reacting. Such reflections can therefore be considered as *narratives* that attempt to construct an order out of potential chaos.[28] They are also ways of considering the important link between people and practice. It needs to be said that what people say and what actually they do—as is obvious not just in farm practice, but also in policy practice—is often at a disjunction. In the case of this evidence, we accepted what people were recounting to us in terms of their own practice.

These narratives have been restructured here at the individual, partnership, family and community levels, and draw out the themes of innovation, accommodating change, building a future and sustaining communities.

Individuals—managing innovatively

There is a strong correlation between the drought of the 1990s and the growth in take-up of information technology as a farm management tool. The research found that, for many families, turning to sophisticated farm management approaches using information technology enabled them to better manage their resources, and this in turn enabled more strategic planning for farm production. One example was the increased use of climate information technology—particularly that of the ABC-TV weather Southern Oscillation Index (SOI) reporting—which became incorporated into everyday practice. Such technologies required a 'rethinking' of the approach to production, and many of those interviewed explained how they had re-educated themselves, developing expertise they had previously not had.

> I thought it was a bit of a joke. Temperature in Darwin and Honolulu, a lot of nonsense. We realise now that while that line is under the bar we are in trouble. I think it will influence our decisions in the future. Sell cattle early in the year, not hang on and hang on (Neil, cattle grazier, CQ).

The SOI was not the only tool utilised. The *Rainman* software package also became an important strategy. Producers were able to compare rainfall rates in the current season with those of years ago, even close to 100 years ago. For example, in Central Queensland, the worse drought was statistically identified as occurring in 1902, until the drought of 1991–1997, which 'came close' to it in many districts. As was discovered through the research, such records became an important component in the struggle to mount a case for assistance.[29]

Information technology also enabled a reorganisation of farm accounting practices, and many took advantage of opportunities to upgrade budgeting requirements and the technologies associated with them.[30] In a number of cases, women were the innovators here, as the bulk of 'bookwork' was their responsibility. In this way, as the research found, women began to take on a more active role in farm management, learning the new technologies in order to contribute to the whole enterprise. Helen and Ed (sheep/wheat farmers) explain the impact of their experience:

> We always did a budget, but we never really understood budget—I shouldn't say we didn't understand, but we just weren't really, totally focused on it. Now we can draw a budget up. We can do a net profit, we've got all our assets, all our liabilities. We know what our sheep earn and how much they

cost us to run per head. Every paddock is listed. With a grain paddock we know what is going in it and what's coming off that paddock, and what that paddock costs. We've done a lot of that (Helen).

... I didn't understand computers and [Helen] didn't understand computers. We didn't understand the system and now we [do] understand the system and it's good. It's been hard to learn (Ed).

... we've only had the computer for four years ... we virtually educated our accountant the way we wanted, and I can pretty much say we've got the bank manager thinking the way we think ... (Helen).

New forms of farm management practices also emerged as strategies developed for difficult times. During the interviews we were told of a variety of alternative approaches that producers had trialed in order to manage the risk better. Such innovative strategies were associated with water usage, animal production and the use of fertilisers. One interesting example involved a management program that challenged all 'known experience' in that district. Paul explains:

...what we've done is split our stock in half so that we [are] basically running two properties on one block. What are the advantages? Income more evenly distributed over the whole year giving a better cash flow, great selection of markets, less brands required so you can select higher grades, only handling half the sheep numbers at any operation and better spread of work load. These days ... with costs increasing all the time, you've got to make every post a winner. You've got to run good stock whether it's cattle or sheep; they've got to be good. I'd recommend it to anybody ...

The use of new technologies and enabling strategies gave individuals incentives to continue in difficult times. Such innovation was not limited to younger producers, as in many interviews even those who had been farming for many years were also keen to try new forms of production. However, the research did find that the growth of 'experts' during the drought often placed additional stress on their individual capacity to cope. As a result '… their relationship with their property became a struggle between those "experts" who provided advice counter to their own perceived experience in managing their properties'.[31] This experience identifies the importance of establishing a balance in 'local' *versus* 'external' knowledge in drought management strategies. In other words, there was a widely held view that individuals were not using new technologies, but relying only on precedence. Thus the research highlights the need for this capacity of individuals to take up new technologies rapidly, for the new technology to be accepted and therefore incorporated into future policy formulation.

Partnerships—accommodating change

As discussed above, families were the first 'line of defence' and the partnership between husband and wife was central to that. In the interviews conducted separately, both spouses spoke of the importance of the other in their own capacity to maintain strength and resiliency. At the core of this relationship was the acceptance of change, particularly changed roles. For many women, the drought meant that more was expected of them, and they were drawn into the farm management in ways that challenged previous, more traditional, approaches. In Central Queensland, for example, where agistment was still possible in the early years of the drought, women were left behind on one property while their husbands moved to another and, as a result, women who had not had this responsibility earlier, became involved in the struggle to feed and water stock. Many of them talked of moving 'outside' the home for the first time since their marriage,[32] and how 'strangely enough when I'm down at the yards I feel better about the drought because you are physically helping with something' (Bronwyn, cattle grazier).

Sharing the stress also became a crucial strategy that enabled resiliency. While at the interviews women were more likely to talk about their own and their husband's stress levels, men were more likely to down play their health responses.[33] In many cases, women described how they 'acted as buffers' against much of the external stress.

All the bookwork comes to here. The bank manager rings up, well, everything's here. The accountant rings here. We get the first line of contact with everyone—usually me' (Bronwyn, cattle grazier).

However, the research did find interesting differences between families in terms of decision-making. When asked if major decisions were made alone, or involved the partner—27 per cent of men stated that they took the decision alone—while 10 per cent of women stated that they were not involved at all in any decision-making. The majority of women—85 per cent—were only partly involved. As the crisis continued, and roles within the marriage changed, women were more likely to become involved in decision-making.

Another major strategy to enable long-term management for many was the capacity to undertake off-farm work. In some areas this was not possible, as employment was not available. However, in cases where it was, both partners spoke of the importance of being able to work off farm,[34] and thus enable the farm production to continue. Another interesting finding was that some women producers turned to on-farm enterprises as strategies to build more income.[35]

The crucial nature of this changed spouse relationship as a buffer against the ongoing crisis came to be slowly recognised by those external agencies already helping, as well as agribusiness, many of whom, initially, were still operating within older, more traditional models. The important role of women in farm management is now well understood, as *FarmBis* and other schemes are targeted at both spouses. Advertising in journals, and marketing by agribusiness has also increasingly recognised the important decision-making role of women. Future policy formulation will neglect the role of women in farm enterprises at its peril.

Families—building a future

Two key issues emerged from the research that enables a type of resiliency in families to be explored. The first highlights the importance of intergenerational planning to the whole farm enterprise. The second focuses on the use of family labour in times of crisis. With the median age of farmers across Australia of 48, and over 55 in some broadacre farming areas, in 1996,[36] discussion of intergenerational planning featured strongly in the interviews.[37]

This discussion also identified the key importance of the extended family members in the capacity of individuals to continue to manage during the crisis. Many named their family members as critical in providing such support, and, contrary to North American studies, there was not much 'blaming' of other family members[38] for the crisis. Many respondents reflected on their own capacity to keep going, and their planning for retirement after the drought was over. However, this was usually reflected on in relation to the management of property in the present crisis, and wanting to establish it so it could be managed well in the future. According to one wheat/sheep farmer, this was to 'to set it up for the family to carry on … [and] make it as drought proof as I possibly can.' This desire to manage the present for the future became a fundamental issue for families. Robert and Barbara, who had three children between the ages of five and nine, discuss this:

> On 50 000 acres you'd have a good lifestyle, but if you wish to make provision to educate your children … (Robert)

> … another thing that is strange about farmers, that [Robert] feels he has to be able to provide that if our son wants to be a farmer it has to be sitting here waiting for him. They feel they have to have it, if the boys want to be farmers, they've got to be able to provide that for them. I can't think of any other business where they think they have to set up and be ready for the boys to take over. It doesn't happen in any other business, but it does in this business … (Barbara)

> I don't think that's strange. (Robert)

> That makes it hard. Everything we are doing we have to get bigger because … it's not going to be viable and that's a big pressure to put on yourself. (Barbara)

The other aspect of the intergenerational issue was increasing use of family labour. As the drought continued, many producers 'managed' by divesting themselves of casual or hired labour. Increasingly, more and more tasks were undertaken only by family members. If there were no family members available, husbands and wives found themselves managing a whole property, taking up work that had been done by many others, by themselves. In addition, as the crisis continued, labour, even if it was possible to afford, became difficult to obtain, as casual labour abandoned agricultural production. This in turn impacted on the sheer hard work associated with managing properties in hard times. Both partners and their extended families had little time for social activity, a fact that impacted directly on their communities (see below). Jean, a cattle grazier, explains:

> *It has made everybody so busy. I'd say that everyone has had
> to shed their labour, hardly anybody I know is employing labour.
> They are just working themselves now and just doing twice
> as much.*

Mary, also a cattle grazier, reflected on the impact on her family. For her the big question about whether to stay or whether to go hinged around the 'family unit, whether we'd be better off battling ourselves and going without and keeping everything together for the children's sake'. However, the family relationship had changed as a result of the decision to stay. 'It has been strengthened, because we all work together and the kids understand what is going on.' Mary's children had become involved in the management of the property, and her daughter 'actually started doing all the accounts. She knows where all the money goes. She is very aware of what happens. They don't tend to ask for things when they know they can't have them.'

These views were shared by many of our respondents. For example, Robert and Barbara were planning for the future, while managing the present.

> *We feel we don't need a lot. We just think that as long as we
> can give the kids the best education we can give them, we can't
> do anything more for them ... we want to be able to provide a
> sound background for the kids and give them everything that we
> possibly can ... (Barbara).*

Policies developed for families in times of crisis, such as drought, should take into account the strength of family ties, and the powerful incentive of intergenerational planning. Planning for the future became a strategic component of many of the farm financial counselling opportunities made available during the mid-1990s.

Sustaining communities

There has been much discussion in the past decade about the sustainability of Australian rural communities. Certainly, we have seen population decline, reduction of services, fewer small businesses, less public transport, bank closures, high unemployment and the challenge of changing agricultural production. In the early part of the 1990s, there were many predictions made about the future of small communities. 'Tiny towns must die' became a common headline[39] and further decline was seen as inevitable. It became hard for communities to fight against what appeared to be an inevitable outcome. For many of those involved in the research, this emerged as a social justice and equity issue as they questioned why government assistance was provided for other disastrous events, but also why assistance was available when people were not in as dire straits. It was also raised as an issue when taxation and the welfare system was discussed. Many male farmers believed that the 'fair go' principle should be adhered to as agriculture kept the country balanced in terms of international trade. Finally, the 'fair go' issue was also raised in regard to the equity of distribution of any assistance and how the whole process of drought declaration then impacts on communities and families.

During the research, people reflected on these consequences of drought, and on the way in which social networks had fragmented as a result of the ongoing crisis. There were comparisons made between previous years and the years of drought, in their own community participation. Many identified the drought as the major reason why they no longer participated as much in such activities that could not be directly associated with farm production. An important finding was that the women we interviewed spoke of not being able to participate in community activities, such as CWA or art groups, as they came to understand that such activities were not contributing to the farm enterprise. Men were not as involved in sporting activities. On the other hand, events associated with Landcare, agricultural field days and animal sales were still considered worth the effort, and the expenditure of diminishing resources. We were told of Rotary and CWA groups that had dissolved due to lack of membership, and this confronts us with the question: whose responsibility is it to pick up such fragmented networks after the crisis has past?

In many communities, this re-establishing of social capital has become the province of human service workers such as community development officers, farm financial counsellors, social workers or, in the case of Queensland, local area coordinators.[40] Such 'social capital builders' are primarily supported by governments, who recognise the fragility of communities, and are prepared to support the reintegration and building of future networks of support. This is underpinned by a philosophy of self-reliance, as it is founded on principles of self-empowerment and potential individual resiliency. Certainly, the crisis showed that people cannot do it alone. They do need others to continue, and, in the same way, neither can individual communities do it alone. They need guidance and support from the Commonwealth and State governments. A 'hands-off' approach does not enable resiliency to re-establish social cohesion. For many communities, the challenge of the crisis created an opportunity, one that has resulted in heartening stories of a turn around for them. Leadership has emerged as a key factor, as has collaboration between key stakeholders—including, importantly, local government.

The research made a number of important findings in relation to the sustainability of rural communities post-crisis. First, that the drought has 'changed the way in which the land is managed and the structure of employment'.[41] This means that a 'return to the old days' is unlikely, and that in-migration of new populations is an important future strategy to enable the economy of small communities to remain viable. Such in-migration needs to be based on an economy that encourages people to return to environments where perhaps previously they had not considered living. In the 1990s there was a growth in 'sea change'-lifestyle decisions, and not all of these seekers for quality of life outside of cities want to live on the coast. Such in-migration needs to be planned for and social networking encouraged, so that the outcome is an integrated community, not one where a 'divide' between old and new is sustained. Second, that the support of human services is likely to remain important, and that for many, the drought assistance provided becomes a key in being able to maintain a sense of purpose and future. There is currently early evidence that such human service (or welfare) support has become 'easier' to call on, and in the case of the current drought, families are not waiting until there is no other option. However, such assistance needs to be provided with sensitivity and understanding about individual, family and community pride, and about the potential for community division associated with inequitable resource distribution. Third, those decisions undertaken in the drought of the 1990s are now impacting directly on our current experience. Whatever we are doing now we are creating our future. This is true not only for farmers but also for policy-makers. *Thus how resiliently we accept the challenges of this current drought has been established in decisions made during the previous one.* Finally, and importantly, the potential for division between 'urban' and 'rural' Australia needs to be well managed within any future policy development. This matter devolves down to the level of community—where local towns feel the impact of declining production on their surrounding farms. However, for many of our respondents, the lack of under-standing about their experience and the hardship they were undergoing was evidenced by what they saw as a 'rejection' by others of their 'whingeing' and 'complaining'.

Conclusion

In this chapter I have provided a 'from-the-ground' perspective of the impact of drought on farm families. I have identified that resiliency—at the individual, partnership, family and community level—becomes a key to managing the crisis and planning for a future beyond it. Drawing on recent, detailed research conducted with farm families in Queensland and New South Wales, while they were experiencing the drought of the 1990s, I have argued for future policy development that takes into account the lived experiences of those affected by drought. In summary, such policy development should take account of the inherent resiliency by considering:

- the capacity of individuals to create innovation solutions
- the importance of new forms of technology and their rapid take up
- the balance between 'expert advice' and on-the-farm experience
- the centrality of the husband/wife partnership
- the changing roles of women in agriculture
- the impact of stress at times of crisis

- the decision-making processes within partnerships
- that intergenerational planning continues to be important as Australia ages
- that family labour is likely to be keeping the farm productive
- that equity is crucial across families and communities
- that social networks become increasingly fragile as the crisis continues
- that social capital building can be achieved through an integrated response
- the demographics of community change
- that pride and self-capacity need to be strengthened.

In addition, some lessons from the research for future policy development include the fact that researchers cannot 'assume the existence of that which may be taken for granted in other quarters'[42]. One obvious example of this is the assumption that social solidarity exists in the bush. This utopian view may exist:

> ... *because of the specific nature of manufactured appearances but, notwithstanding the truth of this proposition, this type of situation points to the importance of researchers acknowledging the informal realm of social structure. This is the often-hidden or out-of-sight domain that lies behind appearances.*[43]

A review of the social construction of drought model (Figure 5.1) can now be suggested as follows:

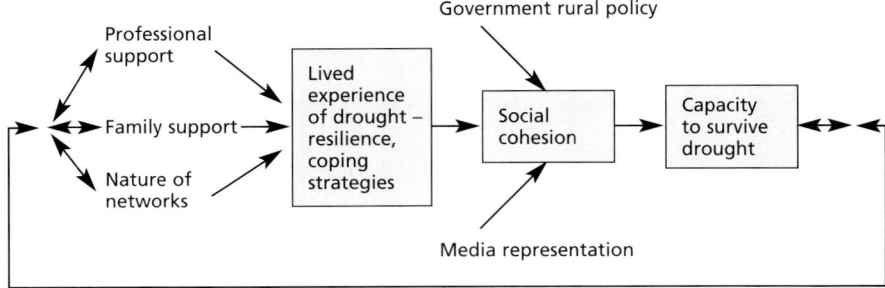

Figure 5.3 Resiliency and drought — a scematic approach

We can see from this model that the issues discussed in this chapter impact on the capacity of individuals to survive drought. Their professional and family supports and the nature of the (extended) networks; the policy that impacts on them in times of crisis and in the 'good times'; the representation of the crisis in the media; and the language used to describe their struggle enable a social cohesion that is drawn from the resiliency of individuals, families and their communities.

Future evidence-based policy development will enable drought management strategies to be more targeted and focused. By keeping the resiliency concept to the fore, such future management strategies will not de-fragment communities further, or put greater strains on families than they are already experiencing. Through such strategies it will be possible to recognise the potential to disempower and the potential to create tension. As we now consider the place of drought in our society, those who spent time with us reflecting on their experiences have something to teach us. They did so because they wanted others to understand their situation, that while they are self-reliant they are also interconnected with their communities and with the wider Australian society. Their future is our future.

1 For a North American perspective, see Keenan and Krannich 1997 and Wilhite 1994; for South Africa, Smith, DI 1993 and Bruwer 1993.

2 Ricks 2002, p. 13.

3 Van Manen 1990.

4 Greig *et al.* 2002.

5 Daly 1994; White *et al.* 1995.

6 Heathcote 1998, p. 396.

7 The National Drought Policy also explicitly recognised the need for increased government support during times of severe or exceptional drought. In that sense the policy was designed to respond to droughts of all intensities, with farmers expected to manage 'normal' drought events, but with government intervention when the situation moved beyond 'normal'.

8 Again as I write this, recent reporting from western Queensland has some properties in drought for 11 seasons, and producers are living on drought assistance to survive. *Statewide*, ABC–TV, Friday 19 July 2003.

9 Simmons 1993, p. 445.

10 One broad example can be seen in the way in which volunteer labour as an altruistic practice, has been incorporated into human service policy in the past two decades.

11 Stehlik *et al.* 1999. This project forms the basis of this chapter. I acknowledge my collaborators on this project, Geoffrey Lawrence (UQ) and Ian Gray (CSU), and the more current collaborative work undertaken with Lesley Chenoweth (UQ), focusing on human service practice and community resiliency.

12 White *et al.* 1995, p. 259, provide a model-based analysis of the severity of droughts, but prediction for length of existing ones remains a future hope.

13 See Heathcote 1998.

14 Drawn from Stehlik *et al.* 1999, p. xi.

15 Gray *et al.* 1998, p. 24.

16 Stewart *et al.* 1996/97, p. 73.

17 Stehlik and Chenoweth 2003 (forthcoming). See also, Fuller 2002.

18 Stehlik *et al.* 1999, p. 94.

19 Stehlik *et al.* 1999, p. 94.

20 Graziers will usually attempt to hold on to their breeding stock so as to prepare for the better season after a drought. Unlike farmers, who can usually plant a crop almost immediately after rain, breeding animals need time to be re-established. In many cases, many, many years of breeding decisions are lost when breeding stock are dispersed. This has become a major issue of discussion in terms of compensation in the current (2002) drought.

21 Stehlik *et al.* 2000.

22 Gray *et al.* 1998.

23 Lawrence *et al.* 1999.

24 Lawrence *et al.* 1997.

25 Bulis *et al.* 1996.

26 Barber forthcoming.

27 Lewins, F. 2002. Comments for review of this chapter.

28 Lewins, F. 2002.

29 Stehlik *et al.* 1999, p. 77.

30 Stehlik *et al.* 1999, p. 86.

31 Stehlik *et al.* 1999, p. 52.

32 See Stehlik *et al.* 2000 for more details.

33 See also Gray and Lawrence 1996.

34 In this case, 'off-farm' is a commonly used term to denote any employment activity that brings income into the household, but is not directly related to the production undertaken on the farm. In many cases, we found women were working as teachers, nurses and human service professionals 'off-farm'. However, we also found that many men and women were turning their farms into sites of alternative production, for example, homestays, B & Bs and tourist sites. These can also been conceptualised as 'off-farm' activities. See Jennings and Stehlik 2000.

35 Jennings and Stehlik 2000.

36 Haberkorn *et al.* 1999, p. 56.

37 As a whole, those interviewed in Queensland were slightly older than those in New South Wales; however, the issue of retirement emerged as a key one throughout all interviews.

38 Stehlik *et al.* 1999, p. 65.

39 See Forth 2000.

40 See Stehlik and Chenoweth forthcoming.

41 Stehlik and Chenoweth forthcoming, p. 71.

42 Lewins, F. 2002. Comments for review of chapter.

43 Lewins, F. 2002. Comments for review of chapter.

Economic aspects of drought and drought policy

*Bruce O'Meagher**

Drought is a normal feature of climate variability experienced to some extent by nearly all countries. With the possible exception of the impacts of our contribution to greenhouse gas emissions on the intensity of climate variability, it is a natural phenomenon over which we humans have little influence and no control. As such, it represents a significant constraint on human activity, principally, but not only agricultural activity.

While we are unable to control the frequency, duration and intensity of drought, its impacts can be influenced through the decisions and actions we take. These responses can have positive and negative outcomes, depending on the choices made. How best to respond to drought therefore raises important policy considerations, particularly for countries like Australia, which experience significant climate variability.

In this chapter I will examine responses to agricultural drought from an economic perspective. A short discussion of salient

* Acknowledgements: The author wishes to thank the other contributors to this volume, John Quiggin, David White, Don Brunker and Michelle Croker for their useful comments on earlier drafts of this chapter. The views expressed are the author's own and do not necessarily reflect those of his employer or referees.

economic characteristics of agricultural drought and its implications for farm businesses and the broader community is followed by a consideration of the role of public policy in effective drought response management. Principles for government intervention to minimise the adverse impacts of drought are proposed and the drought policy of the 1990s and prospective public policy responses are discussed against these principles by way of illustration.

Economic characteristics of agricultural drought

An exogenous economic constraint

From an economic perspective, agricultural drought may be regarded as an exogenous, or external, variable that, through its impacts on hydrological and agronomic systems, imposes significant constraints on production and income possibilities. In turn, these constraints influence regional economic structures and settlement patterns and, ultimately, patterns of national economic development. This is illustrated in the maps in Chapter 2. As made clear in later discussions, our failure to fully recognise the implications of these natural constraints has resulted in significant adverse (but potentially avoidable) economic, social and environmental costs.[1]

Uncertainty and risk

Drought involves a high level of uncertainty and considerable risk, particularly for agricultural producers. Hardaker, Huirne and Anderson define *uncertainty* as imperfect knowledge and *risk* as uncertain consequences, particularly exposure to unfavourable circumstances'.[2] In short, although we can be certain that drought will occur, we are uncertain about the timing of its onset, its duration and hence its consequences.

Our uncertainty stems from several sources, including imperfect knowledge of climate dynamics and drought incidence; interactions between meteorological, hydrological and agronomic systems, and of the precise impacts individual events will have on farm production and income; and likely market-based responses to drought onset, including by other producers (at home and abroad), input suppliers and financial institutions.

Drought impacts and related costs

The impacts of agricultural drought are frequently extensive and multifaceted, particularly for extensive drought episodes. These can include financial, economic, environmental, social and political impacts; related costs can be significant and may be experienced by rural and urban areas alike, and by the nation as a whole.[3]

For farm businesses, the onset of drought is likely to reduce farm production and income, increase input costs and, the longer it lasts, potentially increase farm business vulnerability. Prolonged drought episodes can have adverse economic implications for local and regional non-farm businesses—with the severity of impact likely to increase the less diversified the particular local or regional economy is. Such episodes, particularly where they are geographically widespread, are likely eventually to flow through to other rural sectors (such as forestry and fisheries) and eventually to the economy as a whole. Flow-on impacts can include reduced non-farm production and employment, reduced export income, increased consumer prices and reductions in national income. At the time of writing, a widespread drought is expected to reduce Australian national economic growth by $5.4 billion or 0.7 percentage points during 2002–03.[4]

Appropriate on-farm strategies can reduce drought risk and associated costs. Such strategies can be complex to develop and can themselves represent a significant, ongoing cost. They require that farm businesses invest in climate and drought risk management information, plant and equipment, and planning and decision support systems to ensure that farm production systems are geared to deal with eventual drought onset and its possible longer term implications for productivity. Farm businesses and families also need to be sufficiently insured to withstand the loss of income associated with drought onset and drought recovery, and, in extreme circumstances, to ensure that a business exit strategy is in place.[5]

The broader community may bear significant costs through public-sector funding of initiatives to reduce drought uncertainties, risks and costs. In Australia, such initiatives have included: research to increase our understanding of climate variability and its implications for effective drought risk management, predominantly by the CSIRO, Bureau of Meteorology, Bureau of Rural Sciences, State government agencies and universities;[6] investment in irrigation and other infrastructure aimed at increasing productive potential in drought-prone areas; and provision of farm business and farm family income and other welfare support and non-farm business support to alleviate the adverse impacts of drought.

While such initiatives can contribute to reducing drought-related costs over the longer term, inappropriate public-sector choices, including some not directly associated with drought risk mitigation, may increase longer term costs. 'Misguided optimism' about our capacity to exploit marginal areas has resulted in a long history of 'drought following the plough' in many countries, including Australia.[7] Inappropriate investment in irrigation, soldier and closer settlement schemes, and excessive tree-clearing programs in Australia have reduced agricultural productivity, degraded environmental assets and rendered farm businesses, regional communities and the broader community more vulnerable to drought risk and its associated costs.

Drought can produce winners as well as losers. Drought risk management expenditures by farmers represent incomes to upstream and downstream businesses within the agriculture sector, for example for business and property planning advisers, and to the regional communities in which their farms are located. Even after drought onset, relevant services continue to generate income within regional communities. Importantly, drought-induced price increases may benefit those farmers unaffected by drought—and, perversely, given the fickle nature of rainfall variability, sometimes there are lucky farmers within a predominantly drought-affected region who receive timely, adequate rainfall.

It should also be noted that, although it is perceived predominantly as an agricultural or rural problem in Australia, drought can also have important consequences for urban areas.[8] Water supply and related restrictions during periods of drought can reduce access to dams and streams for recreational purposes, increase water supply charges and reduce water consumption for household purposes. The associated costs of these measures to individuals and to the community can also be significant.

Business risk rather than natural disaster

Finally, while all countries are exposed to drought risk, as Wilhite points out in his chapter, the level of risk (including economic risk) differs from country to country. These differences result fundamentally from the range of climatic and related hydrological and agronomic features of each country. However, the impact of these natural factors can be accentuated by each country's stage of economic development, as well as by the drought risk management decisions of private individuals (notably farmers) and governments. Such differences imply quite different perceptions about the nature of drought and appropriate policy responses.

Benson and Clay distinguish between drought risks in lesser developed economies on the one hand and developed, complex and predominantly market-based economies on the other.[9] For the former, drought events can have potentially devastating consequences because of these countries' relatively heavy reliance on agricultural production for domestic food consumption and for export income. In developed economies, on the other hand, where domestic agricultural production is likely to be relatively less significant for meeting basic food needs and for the economy generally, drought impacts can be significant but rarely if ever threatening to national food security, overall economic stability or to human life.

From an economic perspective, agricultural drought in lesser developed countries is therefore all too often appropriately regarded as a 'natural disaster', albeit often made worse by inappropriate human choices. In more developed economies, however, agricultural drought is more appropriately regarded fundamentally as an ongoing farm business risk rather than as a disaster—notwithstanding that such droughts can have serious adverse environmental consequences. In particular, it needs to be recognised that in developed economies, drought risks can be more readily absorbed by a combination of developed markets capable of allocating resources relatively more efficiently; diversified economic activity capable of absorbing adjustment pressures; and welfare safety nets. Such recognition implies a greater emphasis on private-sector responsibility for drought risk management than has been the case historically in Australia.[10]

Role of the Public Sector

Given the potential extent and significance of the costs involved, it is important from a public policy perspective that agricultural drought uncertainty and risk be minimised, although we need to recognise that it is unlikely that either will ever be entirely eliminated. Pursuit of this objective itself comes at a cost, and raises concerns about who should bear that cost and about appropriate private and public sector responses to ensure that economic and other costs are minimised.

There has been a long and at times controversial history of debate concerning the role of public-sector intervention in meeting these challenges. The remainder of this chapter is focused on the appropriate role for governments in drought risk reduction.

Intervention principles

In general, governments intervene in the private marketplace to achieve a range of economic, social and political objectives because it is judged that these objectives cannot, for whatever reason, be achieved without such intervention. Intervention may be justified where private markets are unable to produce the goods and services that the community requires in an efficient manner or cannot produce them at all. However, unnecessary or inefficient intervention can reduce overall community welfare.

A well-established principle of welfare economics is that perfectly competitive markets will deliver optimal, 'first best' outcomes in terms both of efficiency and overall community welfare.[11] But it is also well understood that, in reality, such markets generally do not exist and that the existence of market imperfections or 'market failures' may provide a basis for government intervention to achieve so-called 'second-best' outcomes. Public policy, including drought policy, is essentially about achieving such 'second-best' outcomes.

However, the existence of market failure of itself does not justify government intervention. Intervention comes at a cost—to the taxpayers who fund it and, where intervention is inefficient or ineffective, to the economy as a whole through the misallocation of resources. While private markets often fail to some degree, governments also frequently fail since, among other reasons, they have to deal with many of the same kinds of constraints, notably information constraints, as private markets. Accordingly, reasonably functioning, competitive private markets, though imperfect, may well deliver better 'second-best' outcomes than would be provided through intervention. General preconditions for effective government intervention from a public policy perspective, include that it can be reasonably established that there is substantive market failure; that the benefits from intervention will outweigh the costs; and that any intervention is itself as efficient and effective as possible.

Box 1[12] outlines sources of market failure commonly identified by economists as providing an arguable, but not necessarily conclusive, basis for intervention.

Box 1   Arguable 'market failures' justifying intervention

Competition and structural failure
- Excessive market power, frequently based on increasing returns to scale.
- Uncertainty regarding existence or distribution of property rights

Incomplete markets where goods and services may be under-provided
- For example, absence of insurance for certain risks.

Public goods
Goods (such as national defence or provision of lighthouses) that the market will not supply or will not supply enough of. Such goods are normally (1) non-rival—it costs nothing for an additional individual to enjoy the benefits of provision of such goods—and (2) non-excludable—it is difficult or impossible to exclude other individuals from consuming the good.

Positive and negative externalities
Here the activities of one person can affect—positively or negatively—the welfare of others.

Information failures
Here there is a lack of relevant information available to market participants to enable effective decision-making.

Merit goods
Goods and services that individuals use are compelled (for example, seat belts) or encouraged (subsidies to the arts) to consume/use.

Box 1	Arguable 'market failures' justifying intervention (continued)

Intervention may also be justified on the grounds that private markets are unable to deliver social equity/welfare (for example, income distribution) or intergenerational equity (moderating the adverse effects of our actions today on the welfare of future generations) objectives.

Intervention may be justified on the basis of more than one market failure since some forms of failure are interrelated. On the other hand, intervention based on one factor may conflict with intervention based on another. Ambiguities and the exercise of policy judgement are therefore involved in many aspects of public-sector decision-making based on market failure considerations.

Cost/benefit analysis provides a useful but rarely decisive tool for assessing whether intervention is likely to result in net benefits. Information constraints mean that the results of cost/benefit analysis are partial and only robust in specific circumstances. In practice therefore, intervention decisions are usually based on informed judgements that take account of the nature of the problem (is there a substantive market failure?) and the likely net benefits (are the benefits of intervention likely to outweigh the costs?). This is particularly so where there is significant ambiguity, including, for example, where there are multiple policy objectives.

In this situation, it is important that public policy decision-making is transparent to enable effective scrutiny, and that it is based on processes that are likely to yield reasonably robust, efficient and effective outcomes. Box 2 provides a guide to considerations that I believe will contribute to an improved approach to public decision-making.

Box 2 Structured decision making for government intervention

The proposed measures should:
- have clear, transparent objectives that address a clear and identifiable problem (usually involving substantive market failure or impediments to efficient resource allocation)
- address the cause rather than the symptoms of the problem

Box 2	Structured decision making for government intervention (continued)

- be feasible and be consistent with higher order policy objectives
- result in an increase in efficient resource allocation and in aggregate net benefits within the affected sector and within the economy as a whole (an overall increase in efficiency).

Proposed and alternative measures should be subjected to cost/benefit analysis to ensure that proposals yield net benefits in the most efficient way; where this is not possible for practical reasons (for example, inadequate information or the likely impact does not justify the cost), qualitative evaluation that provides a reasonable basis for assessing whether the likely benefits exceed the likely cost should be undertaken.

Proposed measures should be able to identify expected outcomes and results, and specify parameters against which performance can be assessed.

Proposed delivery arrangements should identify performance indicators against which effectiveness and efficiency can be measured.

Accordingly, implementation of proposed measures should provide a basis for ongoing monitoring and review.

Is drought intervention justified?

Based on our previous discussion, key strategies for effective drought risk management would include three key elements. First, they should reduce uncertainty about climate variability and its interactions with highly diversified hydrological and agronomic systems, and uncertainty about the likely market responses to drought onset and progression. This can be done principally through research and effective articulation of the management implications of the results of such research. Second, they should reduce drought risks, principally through the development and adoption of: requisite planning and decision tools; income insurance and/or protection strategies; and, for extreme circumstances, contingent exit strategies. Finally, they should provide for social welfare and inter-generational equity objectives.

However, when assessed against market failure criteria outlined in Box 1, few of the risk management requirements outlined above would seem to suggest an overwhelming case for extensive public intervention. Each of these strategies are now considered in turn.

Research

Arguments based on market failure considerations are strongest with respect to drought-related research. Despite rapid improvement over the past couple of decades, our understanding of significant areas in drought risk management remains constrained, including in areas such as the interaction between national and regional climate variability, hydrology and agronomy. Basic research in these areas can generally satisfy non-rivalry and non-excludability tests for public goods and, without public-sector intervention, is likely to be under-provided or not provided at all. Such research also has the potential to result in significant positive externalities or spillovers for the community as a whole by informing private markets and public policy analysts and decision-makers about appropriate risk management strategies. This will also provide basic information sets for the development of effective policies for addressing intergenerational concerns through improved management of the natural resource base.

Market failure is not characteristic of all areas of drought risk management research, however. For example, market failure is less likely in areas of applied research with strong potential for commercialisation. But even here there can be a degree of ambiguity. Research into drought-resistant trees, pastures and crops, where the property rights over genetically engineered species are vested in the hands of a private company rather than in the community, is an example of where public intervention may not be justified. On the other hand, there may be strong arguments favouring intervention in such research programs to ensure public safety. Where ambiguities of this kind arise they are likely to be resolved through the political process, although economic analysis can contribute to evaluating the costs and benefits of different courses of action. They may also result in public–private research partnerships where the resulting intellectual property rewards are appropriately shared.

Information generation/dissemination

There are reasonably strong arguments for government intervention in certain aspects of information provision, including the articulation and dissemination of drought-related research. Such arguments are based principally on information failures resulting from insufficient incentives to sustain fully operational and complete private markets.

Private-sector information markets have grown strongly, however. Coinciding with our improved, but still incomplete, understanding of climate and hydrological and agronomic systems, there has been a rapid expansion of information markets and in the availability of associated planning and decision tools, together with a range of new, integrated farming technologies, including precision farming techniques.[13] Much of the impetus for this growth has come from publicly funded research, information provision and training programs. Nevertheless, a significant and growing proportion of information, planning and decision services is provided by the private sector through consultancy services, banks, agro-chemical companies and commercial partners.[14]

Income insurance/protection

Whereas currency and product futures markets may provide adequate hedging against price fluctuations, income protection for production or yield variation is somewhat more problematic. Much of the literature on drought-related income insurance/protection therefore focuses on the apparent incompleteness of formal insurance markets based on problems associated with adverse selection and moral hazards, insufficient information, inadequate risk pools and high administrative costs.[15] Insurance companies may be discouraged from offering policies covering multiple agricultural risks because their very availability may encourage producers to adopt sub-standard risk management practices,[16] while insuring against meteorological events appears to have been unattractive to Australian insurers based on a combination of the factors noted above.[17] Publicly funded schemes have been seen as appropriate in these circumstances, particularly where farmer contributions reduce the potential moral hazards involved.[18]

Other factors suggest that the case for intervention is not as strong as it may once have been. Improvements in our understanding of climate dynamics, improved databases on climate and production variability, and increased sophistication (and expe-rience) in reinsurance increases the potential for event-specific insurance.[19] However, the potential may not be realised where governments, regardless of formally announced policies, continue to come to the rescue of 'farmers in trouble' due to drought. This is likely to have deterred both the supply and the demand side for formal drought insurance in the past, and will do so in future unless disincentives associated with intervention are removed.

Importantly, there is also a range of alternative income insurance/protection options for drought-affected businesses—including through the build up of liquid financial reserves, bridging arrangements with financial institutions, temporary off-farm employment, diversification, draw-down of off-farm investments and access to the general welfare safety net in extreme circum-stances. Effective farm planning, investment and decision-making throughout the drought cycle are likely to make a powerful contri-bution to moderating income loss.

Exit

There are no substantial barriers to farm exit, although it is arguable that there are limited social equity/welfare arguments for providing government assistance for job retraining and relocation over and above what is available to the broader community. Specific farm-related intervention could, for example, compensate for the relative disadvantage of more isolated farmers forced to leave the farm or to assist farm adjustment on the purely pragmatic basis that it addresses the problem of 'stickiness'—the tendency to hang on and stick it out because the farm is also the family home and there is potential for some farming families to subsist without much, if any, assistance for relatively long periods of time. But once again, a history of intervention may have encouraged such 'stickiness'.

Social welfare and intergenerational concerns

While minimum income and welfare provision are accepted requirements in developed economies, there are potential moral hazards associated with such provision to drought-affected farm families, since it may act as a disincentive to the adoption of appro-priate risk management strategies. It is for these kinds of reasons that Australian authorities, for example, have provided welfare access to farm families (and other small businesses) only in narrowly designated circumstances.

Intergenerational considerations are relevant to several aspects of drought risk management, particularly as they relate to natural resource and habitat protection. Potential reasons for intervention here could include ambiguity in property rights (for example, who owns natural water flows) and negative externalities from inappropriate management regimes (for example, excessive water storage practices by upstream producers damaging downstream wetlands).[20] Though inappropriate drought risk management practices may accentuate intergenerational concerns, the case for intervention is likely to be more appropriately addressed in an environmental protection rather than in the drought-specific context.

Summary

From an economic perspective, therefore, there is no basis for believing that comprehensive public-sector intervention for drought risk management is justified on market failure grounds. Intervention in specific areas, such as drought-related research and information provision, are likely to be justified, but only where private markets are unable to provide relevant services at less cost. Market failure considerations provide little or no basis for intervention to support drought risk management planning and decision-making and contingency exit planning, given the widespread availability of private-sector provision or lack of impediment. While it is arguable that formal private-sector income insurance has been under-provided, government intervention in this area is likely to have created disincentives for market development—alternative forms of income contingency provision, though possibly less efficient, are also available.

Policy during the drought of the 1990s
National Drought Policy
Australian governments have intervened in one form or another in most of the strategic response areas required for effective drought risk management over much of the period since European settlement.[21] Standing, structured support programs have been in place since the 1960s.

For the most part, such intervention has been aimed at assisting farmers to 'drought proof' their properties or following drought onset. Although not discussed in detail here, intervention has also frequently involved infrastructure and other initiatives to make specific regions less drought exposed. These have been pursued with varying degrees of success. Other non-drought initiatives, as noted previously, have also affected drought exposure, often negatively.

Under the National Drought Policy, discussed by Botterill in this volume, drought is no longer considered a natural disaster, but rather it is seen as one of several sources of uncertainty, the associated risks of which are expected to be integrated into normal farm business risk management. Rather than providing support to 'farmers in trouble', government intervention is now intended to provide a supportive environment for increased farmer self-reliance and minimisation of adverse impacts on the natural resource base.

Support is principally intended to assist the transition to self-reliance; support for farmers in financial difficulties is only to be provided during what has become known as drought exceptional circumstances; that is, circumstances regarded as being beyond those that could be reasonably addressed by normal drought risk management. A number of criteria were, and continue to be, used to establish the existence of such circumstances. The key threshold condition for drought exceptional circumstances support in 1994 was that it could be established that an area to which support was extended was actually experiencing 'exceptional circumstances'. This threshold condition was met if it could be shown that the area had experienced a hydrological deficit of at least 12 months duration which constituted a one in twenty to twenty-five year event. While subject to significant farmer criticism, the triggering of Commonwealth support for drought-affected farm businesses was, for the first time, based on reasonably objective, science-based criteria.[22] The threshold criteria have since been progressively watered down.

Taken together, these broad policy objectives represent a significant advance from a public policy perspective. Overall objectives were articulated in a reasonably clear fashion and the broad nature of the problems being addressed was identified, albeit somewhat obliquely. This was understandable, given that the policy was articulated during a severe drought episode.

Box 3   Principal intervention tools under the National Drought Policy in the 1990s drought

Component	Intervention tool
Research and development	
• government agencies	Budget expenditure grants
• competitive	
Water and fodder storage infrastructure* depreciation	Tax concession—accelerated
Information transfer	Budget expenditure
Education and training support	Grants
Property management planning	Grants
Business support	
• productivity support under 'normal' RAS**	Interest rate subsidies
• drought business support under exceptional circumstances provisions	interest rate subsidies
• water, fodder and transport support	Transaction-based subsidies
Income protection (initially through Income Equalisation Deposits and Farm Management Bonds, later Farm Deposit Scheme)	Concessional tax rate
Social welfare support	Welfare payments
Exit support	Grants

* This measure ceased in 2000.
**Rural Adjustment Scheme (RAS) was superseded by Agriculture – Advancing Australia initiatives in 1998.

The policy detail stands up less well against other intervention criteria, however. Expected outcomes were specified in only the broadest terms, and performance criteria to enable assessment of particular initiatives were not specified at all. From an economic perspective, few of the measures adopted (Box 3) address clear market failures; most addressed symptoms rather than underlying causes; and some are inconsistent with overarching National Drought Policy (NDP) objectives.

Of the measures incorporated in the National Drought Policy framework, only research and information transfer arguably addressed unambiguous market failures. Other measures, including education and training, property management incentives and income protection, address areas where there is only limited or ambiguous evidence of market failure. Substantive cost/benefit analysis of the intervention measures was not undertaken.

The continued use of interest rate and transaction-based subsidies as key intervention tools for drought-affected farm business support is of particular concern from an economic perspective. As Freebairn and others have noted over a long period, these measures do not address substantive market failure; they are likely to benefit non-farm more than farm incomes; they tend to reward poorer performing farmers ahead of those who have undertaken effective drought mitigation efforts; they encourage investment in subsidised activities and away from measures that may be more effective in positioning the farmer to deal with ongoing drought conditions and associated income losses; and they encourage higher stocking rates than are likely to be sustainable from either an individual farm business or environmental viewpoint.[23]

Such measures undermine the objectives of the NDP and contribute to the ongoing 'hydro-illogical cycle' posited by Wilhite.[24] They have the potential to reward poor performers at the expense of better managers; they confuse policy messages concerning the need to move towards self-reliance; and they are ultimately self-defeating because they only strengthen expectations and associated political pressures to help 'farmers in trouble' at an early stage in successive drought events.

Conclusion

The foregoing discussion suggests that agricultural drought has important economic characteristics and that an economic perspective is important to determining private- and public-sector responses to drought incidence. Drought may be regarded, from the economic perspective, as an exogenous variable—that is, as a natural phenomenon which we can anticipate but which we cannot control. Put differently: drought is a reality of Australian life to which we have continually to adapt. As an economic factor, drought is regarded as a factor which we are uncertain of and which involves considerable risks, including significant economic costs. In this sense it is very much at the heart of one of the central themes developed in this book; namely, the process of 'learning to be Australian'. This chapter has suggested that while that learning process has advanced rapidly over the past decade or two, it is far from finished and is likely to remain a continuing challenge.

Though we cannot control drought, private and public decisions and actions can influence our capacity to deal with its consequences. Because of the significant economic and other costs involved, it is important to ensure that those choices are the most appropriate possible since, to the extent that they are inappropriate, they add to the costs involved. As discussed, we are still living with the costs of past inappropriate private and public policies; some current policies are adding to those costs, the burden of which will be borne by current and future generations.

An economic perspective is critically important to how we perceive agricultural drought. While drought may appropriately be regarded as a natural disaster in lesser developed countries, this is entirely inappropriate in a developed country like Australia, which has the economic resources to both reduce drought uncertainty and its attendant risks, and to better absorb the economic, social and environmental consequences involved. In countries like Australia, drought is therefore more appropriately regarded as a normal (though significant) private farm business risk.

This is not to say that there is no role or only a minor role for the public sector. It does mean, for efficiency and effectiveness reasons, that public-sector intervention must be well based. This chapter has argued that the concept of market failure combined with intervention/decision rules that promote transparency and efficiency are critical components for effective intervention. When judged against relevant criteria, important aspects of past and current drought intervention are questionable and in some cases clearly inappropriate.

Notwithstanding the complexities and sometimes conflicting objectives involved, future drought policy development will need to have regard to these concerns. The following observations are made as a contribution to possible public policy improvements.

Addressing market failures
Research and information dissemination
While the case for publicly funded drought research remains strong, the opportunity should be taken to review whether private-sector research is unduly constrained by intervention and whether there are opportunities to better integrate public and private research activity. Development of drought-resistant crop and pasture species, for example, could prove to be fertile ground for investigating the benefits of public–private-sector partnerships.

Farm-based observations of local hydrological and agronomic responses to drought events already contribute significantly to our understanding of drought dynamics and form the basis of much farm-based planning and decision-making. While there is considerable personal incentive to continue learning from such observations, there are few incentives, or for that matter, formal avenues for the results to be scientifically verified or shared with neighbours and the broader research community. Options to address this market failure could include government support for the establishment of a national network of regional drought research groups providing a forum for establishing appropriate methodologies for on-farm research and for the dissemination of the results of both public and private research effort. Such a network could also address the reported lagged take-up of research by some farmers. Farmer participation in such a network is only likely to be effective, however, where intervention which encourages moral hazards is removed, and there is a clear and consistent message that self-reliance would underpin such intervention.

Education and training

As noted above, it is important that policy-makers take stock of the extent to which support of education and training initiatives has led to a discrete shift in farmer risk management behaviour. While it is accepted that risk management planning and associated decision-making processes will continue to be important components of publicly funded education and training initiatives, it is important that the role and capacity of private-sector provision be assessed against the background of the rapid growth in decision support services provided by the private sector over the past couple of decades.

Mutual support can play an important role here. The utilisation of the farmer-based network referred to above could also be considered.

Income insurance/protection

Income insurance options for farmers have expanded, but the formal insurance market remains somewhat immature. The current Farm Management Deposit Scheme has the positive characteristic that it is based on farmer contributions, but it does not entirely eliminate moral hazard and is likely to have constrained the emergence of widespread privately provided insurance. Two possible options to support market growth involve transitional arrangements that would see a phased withdrawal of government intervention in this area. The first involves the continuation of the Farm Management Deposit Scheme for a set period (of say five years), combined with a scaling down of the concessionality of the scheme; the other involves a faster phasing down of the subsidy with a view to privatising the scheme. A pre-announced with-drawal date would have the benefit of signalling to farmers that greater self-reliance would be expected and to potential private-sector service-providers that market opportunities would emerge.

Exit support
Current farm exit policies include counselling, a re-establishment grant and limited cash grants for retraining assistance for farmers to gain skills that would enhance their prospects for employment after farming. However, it is interesting to note that state officials have reported to me that there is evidence of farmers utilising the support that is available to move back into farming at a different location.

Publicly funded relocation and retraining assistance is usually provided only in the context of special, time-limited sectoral adjustment packages, including as a feature of specific agricultural adjustment packages. There are therefore few equity based arguments favouring standing provision of such support, including during drought. Such programs are not provided to mining sector employees made redundant by a downturn in international commodity prices, for example.

Consideration should be given to phasing out this support. Alternatively, consideration could be given to providing more specific adjustment support. Options would include replacing the re-establishment grant by repayable Higher Education Contribution-type loans, or by a combination of location-specific relocation assistance and retraining vouchers that could be used over a fixed timeframe of, say, five years.

Welfare support
There is an arguable case for the continuation of welfare provision to those adversely affected by extreme drought events. But there is

also a need to ensure that such provision is not used to prop up inappropriate farm management. Difficult judgements are involved here since the dividing line between providing welfare safety net support and de facto business support is often a thin one, made the more difficult by the fact that the farm is so often both business asset and family home. Equity considerations are also important, since the owners of other small to medium-sized family businesses generally do not have access to welfare support that can also contribute to the survival of the business.

In these circumstances, there is a strong case for retaining a tough trigger for welfare support. Such a trigger should include an objective, science-based threshold condition for extreme drought to ensure that incentives for drought risk management are not unduly undermined. Alternatively, the possibility of trading off the retention of the Farm Management Deposit Scheme against the elimination of welfare support, except where the farm business is being sold, could be assessed. Such an assessment would, at the end of the day, focus on the relative costs and moral hazards associated with both support mechanisms and the extent to which they jointly or separately undermine the overarching objective of self-reliance.

Addressing intervention measures

Given that the NDP has been in operation for over a decade now, a comprehensive review of its overarching objectives and of how the policy has performed against its broad policy objectives would be appropriate. Such a review should encompass substantive cost/benefit assessment of the contribution and continuing relevance of individual policy measures. Where these are renewed or alternative measures are adopted, they should address clear market failures and have embedded performance indicators against clear objectives so that their policy effectiveness can be reviewed within a predetermined timeline. This last point is important because, as Simmons's 1993 review of the NDP implies, unless there is a discrete improvement in farmers' risk management performance, the 'increased preparedness' approach may actually result in less efficient outcomes than the 'band aid' approach it replaced.[25] The opportunity could also be taken to review the extent to which there is effective integration between internal NDP policies and consistency with other non-drought specific policies having an impact on drought management.

An obvious starting point would involve a review of those interventions, such as interest rate and transaction-based subsidies, which can already be shown to lead to inefficient, counterproductive outcomes. Assessed against the criteria outlined in this chapter, such measures should be eliminated. Moreover, since they do not address well-defined market failures, they do not need to be replaced by alternative measures.

Finally, the linking of the farmer-based networks referred to earlier into a national forum could contribute to strengthening the 'national' aspect of the National Drought Policy. A national drought advisory body could provide the focus for a more integrated approach to setting, or at least contributing, to the establishment of drought-related research priorities and to ensuring it provided a focus for greater inter-disciplinary research. To the extent that its research network provided a basis for meaningful farmer participation, such a body could have the added benefit of helping to break the 'hydro-illogical' cycle by reducing political pressures on government to provide ad hoc drought support.

---

1 Our discussion is confined to the period since European settlement and does not consider the impacts of indigenous land management activity.

2 Hardaker *et al.* 1997, p. 5.

3 See Wilhite this volume. Also Wilhite and Vanyarkho 2000; Benson and Clay 2000, p. 293 chart possible transmission mechanisms of drought shocks through the national economy.

4 ABARE 2002.

5 Krause 1995, chapter 4.

6 Munro and Leslie 1997; Thompson *et al.* 1996

7 Heathcote 1994.

8 Wilhite and Vanyarkho 2000. For a comparison of perspectives in Australia and the United States, see White *et al.* 2001.

9 Benson and Clay 2000, pp. 291–2.

10 O'Meagher *et al.* 2000, pp. 118–20.

11 Refer for example, Rosen 1999, chapter 4.

12 Rosen 1999, pp. 49–53, chapters 5, 6 and 8; Stiglitz 1988, chapter 3; Boadway and Wildasin 1984, chapter 3.

13 Whitmore 2000.

14 DPIE 1996, pp. 22–6.

15 Quiggin *et al.* 1993.

16 Quiggin *et al.* 1993.

17 Mayers 1995.

18 Quiggin *et al.* 1993.

19 Mayers 1995.

20 ABARE 2001.

21 O'Meagher *et al.* 2000.

22 White and Karssies 1997.

23 Freebairn 1983, pp. 185–99.

24 Wilhite 1993.

25 Simmons 1993, pp. 449–50.

Linking environments, decision-making and policy in handling climatic variability

Mark Stafford Smith*

While it is possible to debate separately the biophysical elements that create drought, the social factors that structure producers' experiences and responses to it, and the policy environment that helps define those experiences and responses, ultimately the interactions between all these elements must drive the development of better drought policy in the future.[1] In this chapter I aim to synthesise some understandings developed from the use of coupled biophysical and pastoral decision-making models (HerdEcon, HerdGrasp and RISKHerd) from many interviews with pastoralists, and from naive observation of changing drought policy over 10 years. I seek to do this by describing the implications of evolving human–environment systems and then discussing three case studies of the links between policy and pastoral decision-making. I conclude with a discussion of the implications for future drought policy.

* Acknowledgements: I am grateful to Linda Botterill and Melanie Fisher for creating the impetus for me to think about these issues, and to Bruce O'Meagher for some thoughtful comments on the manuscript. Many people also contributed to the case studies cited herein.

There is a diversity of environments in Australia in biophysical, spatial and temporal terms, as well as in how these features interact with management activities and drivers. It should be noted therefore that the experiences on which this chapter is based are drawn mainly from rangelands and grazing, recalling the profile drought policy has always had in these lands.

What are the special features of the system we are looking at? As described in chapters 1 and 3, all Australian agriculture has to cope to a greater or lesser degree with a highly variable environment over time. This environment is usually characterised by large spatial scales in which it is hard to monitor change, agro-ecosystems with long feedback times, and a diversity of systems within one national jurisdiction. In most areas, population is relatively sparse, with a mixture of very high turnover in some locations and very slow turnover in others. Markets are generally relatively remote and costs of labour and transport high. Communications have been poor, but have taken a radical change for the better in the past decade, and this trend promises to continue for the foreseeable future.

Given these features and the past fixation on central government helping out farmers in times of drought, what lessons have we learned in terms of how to do this and meet our other goals of ecologically sustainable development?

The implications of evolving human–environment systems

Farm enterprises are complex adaptive systems, linking ecology to property decision-making and economic outcomes in the context of changing markets and policies, each element with feedback to the others. In this section, I want to briefly outline some ideas about adaptive institutions that are emerging from the complex adaptive systems, resilience and desertification literatures. This theory is couched very much around issues of relevance to institutions in a variable environment and their capacity to deal with internal change and external shocks without becoming dysfunctional. These are all concerns that ought to inform our planning of future institutional arrangements for agriculture in Australia.

The concept of a complex adaptive system (CAS) is far more

than the sum of the individual words.[2] It is specifically a system in which internally self-organising processes shape the whole system outcomes in ways that are not readily determined from a cause and effect analysis of the whole system. Components of the system change in response to external forces, but also co-evolve internally in response to other components of the system. This occurs in ways that are affected by the middle order number problem (for example, chance presence or absence of significant individuals) and by unexpected events and responses (for example, reorganisation of social groupings) that are not readily predicted from outside. Nonetheless, there is a growing body of understanding that can provide us with guidelines as to when such a system is likely to function resiliently, and when it is liable to dysfunction.

For example, a pastoral property is a complex adaptive system that is based on a (possibly changing) ecology, mediated through a changing property decision-making process that includes family dynamics in the context of a continuously changing market and policy environment, each element with feedback to the others.[3] This complex adaptive system is embedded in a regional CAS involving other land users, industry and other non-government bodies, and a hierarchy of policy-making agencies. The integrated human–environment system evolves in all its elements:

- the state of its physical environment, and their physical drivers (climate, technology, and so on) at a variety of scales (seasonal, between years, over decades)

- local human decision-making in natural resource management policies and reactions to external information.

- policy instruments from all tiers of government (and non-government representative bodies), including the reactions of policy-makers to how producers end up reacting to the policy instruments.

Regional outcomes in pastoral communities and the environments on which they depend are actually mediated by a whole series of system elements that co-evolve, for better or worse. If one accepts the concept of agriculture forming a hierarchy of complex adaptive systems, then one asks rather different questions about the best ways in which higher levels of organisation or government can intervene. For example, if one's view is that pastoralists set stocking rates purely on economic rationalist grounds in the same way across regions and over time, and that this can be managed by external policy settings, then to change the behaviour one simply adjusts the market signals. However, if one is convinced that an array of other social, cultural and environmental issues dominate in determining pastoralists' behaviour, then quite a different approach is likely to be preferred, attempting to provide that regional community with the capacity to plan and adapt its collective behaviour with much greater local sensitivity and appropriate feedback mechanisms.

The new desertification literature is beginning to emphasise the co-evolution of human and environment systems across all scales.[4] In terms of resilience and a complex adaptive systems view, then, the focus is shifting to ask what preconditions will facilitate effective co-evolution. In a further development of the ideas around adaptive management and action learning cycles, it recognises that effective evolution depends on being able to detect and attribute the effects of one's actions, whether one is an individual farmer or a community group or government body, and then on having the knowledge, motivation and capacity to implement any necessary changes to deal with these impacts (see Figure 7.1).

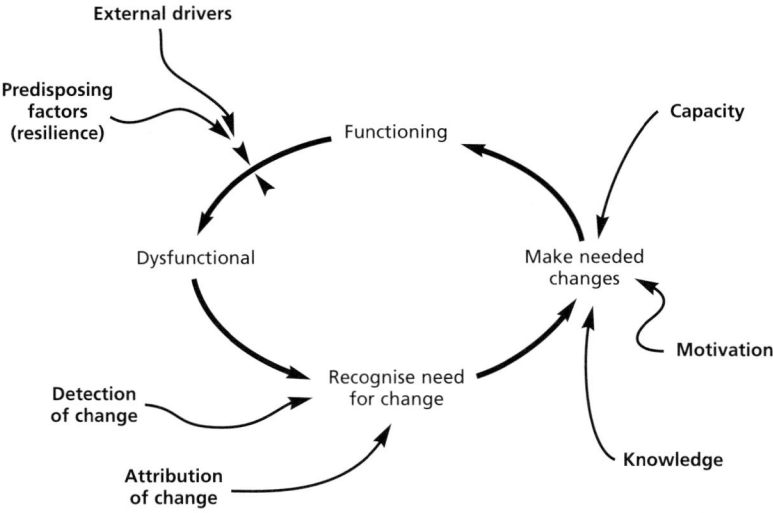

Figure 7.1 One schematic of an adaptive learning cycle (some steps have been omitted) focusing on organisational learning

As noted in Chapter 1, the resilience literature[5] emphasises that, in general, the factors which predispose a functioning institution to fail are the 'slow' variables that render the institution unable to deal with changing externalities. However, it can recover from that dysfunction providing it is able to detect and attribute, and then act on change. These stages can be significantly facilitated by appropriate information and feedback mechanisms, but can also be debilitated by inappropriate contextual signals. This context is generally set by other institutions at another scale, particularly peer groups, industry bodies, government agencies and the policy environment in the case of producers.

Although the critical factors determining the resilience of a system are its 'slow' variables, our day-to-day experience of those systems is usually through 'fast' variables. For example, pastoralists (or their animals) observe day-to-day forage growth, but the resilience of the pasture is more fundamentally driven by slower changing factors such as soil nutrients and water holding capacity, or indeed the species complement of pasture plants that are available. Equally, in the socioeconomic sphere, the day-to-day measures of economic welfare may be one's current bank balance, whereas the underlying critical variables are levels of debt and equity, and even interest rates. In seeking to support long-term change towards sustainable management, it is essential that we focus on those slow variables that set the context, both biophysical and socioeconomic. There are many fast variables, such as grain yield, food reserves and cash flow that humans depend upon in their day-to-day lives; while these are very real issues for short-term welfare, they tend to confuse the strategic debate about coping with drought.

These concepts are particularly important in the variable Australian environment. Where environments are stable, the fast variables are often good indicators of the state of the slow ones (pasture growth reflects soil condition). Where drivers like rainfall or market prices are highly variable, the fast variables merely reflect this variable noise (pasture growth is dominated by rainfall) and it is essential to find a better indicator of the state of the slow variables in the system (in this case, the slow loss of ability to respond maximally to rain when it does come). Resilience theory also emphasises the goal of attempting to expand the domain of stability; that is, the diversity of conditions under which the system can function effectively without passing over some threshold into dysfunction.

One further concept emerges from recent common property resources literature. The literature arose from the debate on the tragedy of the commons, originally applied to communal versus private tenure, but in truth is relevant to many more general governance issues. Work, most notably by Ostrom and colleagues,[6] has begun to tease out the rules by which institutions are constructed, and hence to identify ways of creating these rules that seem to be most effective in different circumstances. One of the most important insights to arise from a melding of this work with the resilience literature is that, where management problems are complex in terms of the relationships to be resolved (as they almost always are in natural resource management), the best institutional arrangements often arise from self-organisation among the players at the appropriate scale, providing they operate in a context that facilitates equitable interactions. However, if every locality or region has a different arrangement, it is still necessary to interface in some common fashion to the next scale up if there is to be any wider coordination (which can usually be shown to be beneficial for certain issues). This leads to seeking the minimum necessary set of characteristics by which the two levels in the hierarchy can interact effectively, and encouraging the higher level to leave the lower level to develop its own solutions within those parameters. This is a very different model to prescribing the entire function of Landcare groups in all areas of the nation, or imposing a standard set of tax instruments universally.

Taken together, all these concepts, which are expanded upon in much greater depth in the various references,[7] highlight a set of issues to which we need to pay attention in considering new institutional arrangements in relation to drought, the pre-eminent symbol of temporal variability, in Australia.

- We need to recognise here, as anywhere, that these self-organising, complex adaptive systems, whether individual pastoral operations, regional communities or larger industry groupings, will function most effectively when they are readily able to detect and attribute the effects of their own actions. Instruments that reduce this ability reduce the resilience of the whole system.

- We must realise that in the environments of variable Australia, this detection is generally difficult, and must focus on the deeper drivers rather than superficial measures of function. Institutions that facilitate appropriate monitoring support resilience; instruments that focus on inappropriate fast indicators distract people from the critical issues.

- Environments and people are changing, and institutions and instruments that are not flexible enough to change with them (or that hold the seeds of stagnation in their structure) are bound to become moribund and create barriers to needed change.

- Given the need for learning and adaptation to occur at the scale of the relevant human/environment interactions, institutional responses need to be carefully matched in scale to the purpose they are approaching. Local knowledge is truly valid in this system, though not as universal gospel; it can interact with a more formal scientific approach to their mutual benefit, in which science helps validate and generalise the local knowledge, and the local knowledge makes the science applicable locally (as well as often being the source of most new ideas!).

In the past, traditional societies had the luxury of many generations to bed down approaches to their environment, which seemed to be 'in balance' over millennia (albeit no doubt won with hard sacrifices or failures over those generations). Today, though, change is upon us at rates that are increasing in an unprecedented way. New ways of developing local knowledge (perhaps depending on appropriate alliances with the scientific method) are required for us to keep up as a society, at both management and policy levels. Again, this is particularly problematic in an environment where it is hard to detect change and cause and effect. If external change occurs faster than our learning, then environmental damage is likely to result; if the environment changes faster than our learning, then our enterprises will suffer.

Importantly, as we come to understand the links between physical, social and policy environments better, we need to recognise that Australian farming systems exist in a variety of different combinations of these environments, with implications for policy. These include all combinations of higher and lower productivity, higher and lower predictability, and nearer and further from markets and the centres of power. Although we pay lip-service to the diversity of our continent, we still generally apply policy in a broadly blanket fashion across these different types of regions. We may need to be more explicit about a classification of different socioecological environments in Australia, aimed at their implications for policy effectiveness.

Case studies of policy-pastoral decision-making analyses

The following three case studies cover a narrow aspect of the whole issue of institutional arrangements in relation to drought, but provide some pointers to broader lessons.

All three of the following studies are generalised on the basis of modelling outputs, but were founded in extensive local consultation or active involvement of producers. This participation had begun during work with the modelling tool RANGEPACK HerdEcon in previous years,[8] and through the DroughtPlan Project around 1993,[9] all of which had a major focus on self-reliance in managing for climatic variability in the short and long term.

Setting criteria for extreme events

In 1996, a project was undertaken to analyse how the national policy-makers administering the exceptional circumstances program (see Chapter 3) might use different measures to act as drought declaration and revocation indicators in an objective system of declaring the worst one-year-in-twenty as an exceptional event. Greg McKeon and I carried out an analysis in rangeland environments. We took existing integrated models of the biophysical and economic aspects of a realistic pastoral property in two different regions, put significant effort into validating them for the drought end of the climatic spectrum, and then ran them over the past 110 years of climate records to examine how various different indices would have performed over that period. The details are described in Stafford Smith and McKeon,[10] from which I draw some key lessons here.

Our underlying thesis was that, on the one hand, the physical measures such as rainfall were likely to be the most objective measures, while measures such as animal production and cash flow would be closer to the hardship that pastoralists were actually experiencing. We therefore simulated a series of indices including rainfall, soil moisture, plant growth, animal live weight gain or wool production, and cash flow. We looked at a cattle enterprise in the Charters Towers region of north-east Queensland, and a sheep enterprise in the mulga lands of south-west Queensland. These two regions and enterprise types represent different conditions in terms of climatic variability, degradation risk and market flexibility. We did not explore the full spectrum of issues in all of these factors, but the two case studies were sufficient to demonstrate how a single policy instrument would play out quite differently in different local conditions.

Our initial results showed how the use of different measures, from rainfall through to cash flow, would have highlighted somewhat different occasions when producers would have been perceived to have been in the poorest 5 per cent of years. Importantly, the major droughts were identified similarly through all measures, but it was the more contentious marginal droughts that varied, suggesting great potential for mistakes in these hard-to-judge events that are exactly the ones for which an objective system is most needed. Thus the first finding was that the choice of index would alter the timing of drought declaration.

Second, as expected, while measures such as animal production or even plant production were much closer to the producer's experience of hardship in terms of low cash flows, these were highly dependent on assumptions about soil types and pasture condition, which are quite specific to individual properties, whereas measures such as rainfall or soil moisture, which could be calculated in a more universal fashion, were less closely aligned with the experience of hardship.

Third, a comparison between case studies showed that the timing and pattern of likely drought declarations also differed between regions and enterprises; one region tended to have more, shorter droughts than the other. In the light of these findings, warning bells rang in terms of how easy it would be to remain objective with any measure. In particular, the results show that a single national measure of extreme drought is likely to create inequity between regions. In particular, they suggested that a single national measure of extreme drought would create arguments between regions in terms of implementation equity.

The greatest insights from the modelling, however, actually flowed from exploring different ways of defining when one would make and revoke a drought declaration, an issue that we had not fully anticipated in taking on the project. How long should a region be in drought before it is regarded as serious? And should the seriousness be judged against the manager's life experience or the full climatic record (which is itself only a snapshot of the long term anyway)? This question becomes particularly relevant if climate change is imposing a trend on the climate such that the future may not be appropriately judged on the past. Thus we explored declarations based on 12, 24 or more months of a measure being in the bottom 5 per cent of conditions experienced during the whole 100-year record, but we also tried allowing the baseline from which the percentiles were calculated to vary according to a moving window of 20 to 50 years prior to the year under consideration. Using the moving window in many ways seems to make the declarations more sensitive and appropriate to the experience of the producers, but it introduced another issue for contention in establishing a supposedly objective criterion for drought declarations—what should the baseline be?

This led us to think further about the criteria by which drought declarations would be revoked (that is, recognising the period when producers are actually in the recovery phase). Because of the recovery times involved, it would not be socially acceptable to revoke a drought declaration, and consequently government assistance, in less than nine to 12 months. The immediate result of this is that, even if one is rigorous about declaring drought in only one in 20 years, the total time spent drought declared must exceed 5 per cent. The amount of this excess depends on whether droughts tend to come in a few big events or many small ones, thus differing between regions. Indeed, if, with the benefit of hindsight, one adjusted the drought entry criterion such that the actual period spent in drought declarations was genuinely 5 per cent of the 100-year record, then the entry criteria would become about one in 25 years, instead of the one-in-20 criterion in the National Drought Policy. This change also would differ between regions for the reasons given earlier. In our paper we explored a number of other permutations on these sorts of issues,[11] highlighting the fact that even with a genuinely objective criterion, such as rainfall, there are still endless options when it comes to the details of how to use it as an index for drought declarations.

What can we learn from this modelling study? The lessons from the point of view of long-term policy goals seem clear to me. First, as stated as far back as submissions to the National Drought Policy review in 1990, in a strict scientific sense it appears impossible to reconcile true objectivity with equity. Thus, the rainfall criterion upsets producers because well-managed properties that have not experienced past degradation still have plenty of feed on the ground at times when other properties are blowing dust. Declaring drought on this criterion is bound not to satisfy everyone. But if one uses a criterion such as grass growth, which is more closely aligned to the conditions of individual properties, and in that sense more equitable in terms of being linked to that real experience of hardship, then the indicator needs to be tailored to every region, if not every property, and is unrealistic to administer. Even with this more sensitive index there are still arguments about attribution, since a good manager on a historically degraded property may be performing similarly to a very poor manager on a property previously in good condition. Second, as science cannot resolve the balance between objectivity and equity, judgments must be made by policy-makers that inevitably introduce a degree of subjectivity into the process of defining declaration and revocation criteria. And third, the way in which any of these 'solutions' play out differs dramatically between regions and enterprises in a way that high-lights the mismatching of the scale of this type of policy instrument with the ecological and social processes that it is intending to influence.

Having said all this, the efforts to bring more objectivity to drought policy in the 1990s were extremely worthwhile. In discussions at the time, Bruce O'Meagher made the strong point that there was a clear educational purpose behind developing a pseudo-objective criterion for drought declarations. This was to raise the awareness of policy-makers that it was illogical to talk of 'exceptional circumstances' and then declare them in more years than not. This is a cogent argument from the point of view of tran-sitory institutional learning, to which I return below. Furthermore, the scientists and policy-makers of the time attempted to work around many of the above problems by creating a basket of indices and by recognising that these were only one input into what was ultimately a semi-political decision-making process about drought declarations. And there is no doubt that for a while this process did serve the institutional educational purposes that Bruce referred to. In the long run, however, it must be said that the policy foundered on the predictable (and predicted) inability to create a truly objective and indisputable measure of drought.

Establishing the relative significance of policy instruments versus price and productivity changes

My second case study is much shorter, with two simple lessons. It was another modelling study founded on consultations with industry in four regions of northern Australia. The project, called Land Use Change in Northern Australia (LUCNA), sought to examine the relative significance of changes in underlying primary productivity compared to changes in markets or policy influences on typical pastoral properties in four regions during 1996–98. It was primarily focused on understanding the potential impacts of long-term climate change on these regions in the context of the other shorter term factors that producers must deal with, but here I want to focus on findings unrelated to climate change.

For this project we spent a great deal of time talking to pastoralists in each region and drawing up a general picture of the influences on their management.[12] We then applied an integrated biophysical and economic modelling framework to simulate typical properties in each region in a similar fashion to the previous drought declarations study. This time we explored the effects of changes in the pastoral productivity, rainfall and temperature regimes, the value of stock purchases and sales, and various types of policy-related financial imposts. We then ran all of these scenarios in combination through 100 ten-year climatic sequences sampled randomly from the real climate record to provide a much larger set of experiences. For all combinations, we examined a series of different management strategies and tactics in order to ascertain what adaptive changes in management behaviour would be expected under the different circumstances imposed by the external drivers.

The key lessons I wish to draw from this case are simple. First, the effects of prices, and then changes in pastoral productivity, far outweigh those of non-regulatory policy instruments. One must be modest in imagining that even major changes in policy-driven financial incentives or costs can play a large role in modifying management behaviour, given the other forces acting on managers. Second, even this limited impact differs greatly among the regions studied. In the same way that market access is often more significant than market premiums as a goad, regulation may be more effective than financial incentives. However, regulation must be carefully targeted, which is hard to do from the national scale for local management on local ecosystems.

The scale of policy intervention must be matched to social and ecological processes in time, as well as matching in space. It is also true that disincentives to act seem to have a greater impact in our human psychology than do positive incentives, so the small impacts here do not mean instruments that send negative messages, however small, are unimportant.

The effectiveness of different tax instruments related to climate variability

The recently completed RISKHerd project aimed to examine the degree to which a number of different taxation provisions support the overarching government goal of moving towards ecologically sustainable development.[13] These instruments, all of which are related to the management of climate variability in agriculture and particularly of rangelands, include a variety of *livestock valuation* methods (determining how stock enter tax accounts in terms of capital changes at the end of a financial year), *livestock elections* (whereby producers can delay bringing the proceeds of drought sales into their declared income), *farm management deposits* (FMDs, whereby producers can set profits aside from their taxable accounts), and *income averaging* (whereby producers apply the average rate of taxation that they would have been liable for in the previous five years) in the context of different *commercial structures* (including private individuals, partnerships and companies). Unlike the previous two studies, RISKHerd explicitly focused on trying to quantify the natural resource ramifications, with the aim of understanding the trade-offs between short- and long-term bottom line benefits—in short, with the aim of putting both post-tax returns and indicators of pasture sustainability (such as soil loss) on the same graph.

RISKHerd was directed by a national steering committee, including the National Farmers' Federation, the Australian Taxation Office, Commonwealth Treasury and Agriculture, Fisheries and Forestry—Australia, with some State department involvement during 1999–2001. Under their guidance we identified a series of current and future feasible instrument combinations. We then surveyed the use of the current instruments in six regions across the rangelands, including sheep and cattle enterprises, and a variety of levels of climatic variability and primary productivity. We also visited each region and discussed instrument use with numerous producers and then, based on these producer consultations, for each region we again constructed realistic enterprises. We ran these through an improved linked biophysical and economic herd model, which this time was further linked to a comprehensive accounting treatment of pastoral income and a full taxation analysis. For each region we assessed a number of management strategies and stocking rate targets, simulating 100 forty-year climatic sequences for each management scenario, and then examined the effects of stochastically varying prices on top of this for a large number (several hundred) of combinations of taxation instrument scenarios.[14]

The full suite of results thus provided a picture of how well each management strategy and taxation instrument scenario performed in response to a contextual environment that was variable in both biophysical and economic domains. Performance was judged on numerous criteria, particularly including profits and disposable income after-tax, the variability of these measures over time, and a number of measures related to biophysical sustainability, such as the mean total standing dry matter that resulted under different management strategies. We were thus able to judge to what extent financial benefits of different instrument arrangements were accompanied by increases or decreases in the likelihood of environmental damage, and make some assessments of the relative trade-offs in these two factors, in the context of a stochastic dominance risk analysis where this was appropriate.

Our first general conclusion was that herd valuation methods based on undervaluing the current capital value of the stock (effectively deferring a substantial tax liability for several years to the time of sale) are financially beneficial to pastoralists and consequently are almost universally used at present. However, these significantly enhance the likelihood of producers holding on to stock into a dry period and effectively increase the risk of degradation on most properties. The significance of this depends on the soil and pasture types, being lower on the resilient black soils and higher on some other soils. Interestingly, although the scale of the financial benefits was even less on average than those tested in the previous case study, in our surveys producers were adamant that their decision-making was affected by this, probably because the impact is in a few specific drought years when large sales are made and is not really averaged out over time.

The second point was that the difficulties introduced by using this advantageous herd valuation method are partly addressed by two other instruments, income averaging and livestock elections, although the former also has numerous other effects. Livestock elections provided little net financial benefit on average, but did reduce the impact of undervalued stock on tax liabilities in years when substantial de-stocking was needed. However, in conjunction with the nominal livestock valuation methods, they tended to significantly increase degradation risk. If livestock valuations became more realistic, livestock elections would become largely redundant for these enterprises. Even income averaging (which costs the public purse an estimated $60 million per year in lost tax revenue and provides significant financial benefits to producers) would be needed less, although interestingly we found no evidence that income averaging *per se* promoted higher stocking rates, unlike the other two instruments, a point I will return to below. In short, one dubious instrument has required a second one to offset its financial impacts in poor years, and *both* are tending to contribute to higher stocking pressures.

Third, if livestock valuation was magically increased to market value, and livestock elections made redundant, there would be a cost in lost short-term profit to the pastoralist. In two regions we were able to make an approximate assessment of how this cost compares to the financial value to the pastoralist of the lost productivity that the higher stocking rates may cause in the longer term. Even at private discount rates, the benefits where we could assess them are similar to the profit lost on a 20-year net present value. At lower, public discount rates the net benefits of changing the instruments are much higher. However, this ignores the costs of making the transition, which are high (particularly for cattle) and beyond the ability of most enterprises to absorb. Thus there is an issue of the extent to which the public good would be served by subsidising this transition; this is considered in the project but is not part of our debate here, other than to note that, as usual, the process of transition is as important a consideration in any institutional change as the final outcome.

Fourth, Farm Management Deposits appear to provide modest benefits to producers, but without promoting problematic changes in stocking rates. As noted above, this was also true of income averaging, although income averaging is not always beneficial (contrary to what its name suggests, it smooths tax paid somewhat, but therefore actually *increases* income variability in most cases). Why is this, compared with the other instruments—that is, what makes a 'good' financial support instrument compared to a 'bad' one in terms of its ancillary effects on natural resource management, and the resilience of the linked human-environment system? The keys to this critical question seem to lie in timing, scale and purpose in relation to management, and bear further investigation at the theoretical level.

The deferral of tax due to low livestock valuations and the use of livestock elections both focus specifically on times when de-stocking should take place; the first discourages de-stocking, while the second makes it easier to do so tardily. Thus both provide incentives to capture short-term financial benefits instead of promoting long-term planning. Income averaging probably really has the same sort of effect, but does not alter resource management (at least in our study) because it is actually not as beneficial as it superficially seems—it certainly increases mean income after tax, but also increases its variability so that, unintentionally, it does not decrease producer financial risk at the cost of environmental risk in the way that low livestock valuations and livestock elections do. Finally, Farm Management Deposits may have been inadequately modelled in our study, since they are much more flexible and hard to develop simulation decision rules for. However, they also provided modest benefits. It may also be that they provide benefits to producers which are much more in tune with the management system, since they effectively shift the decision-making sphere about income variability while leaving it in the producer's hands, rather than imposing a rote solution from outside like the other instruments. If this interpretation is correct, they do not result in such a de-coupling of the financial effects of pastoralists' decisions from their natural resource management implications. In other words, the question is whether the 'subsidy' influences factors that can be passed on to land management or not, and, if so, whether it interrupts the free feedback of information about impacts.

This last discussion requires further thinking, since it could helpfully inform future instrument design. However, if correct, it reinforces the need for policy instruments to improve the capacity of the rest of the system (here the pastoralists, primarily) to obtain the feedback that is needed to test whether their decisions are good or not. The corollary is to ask what features of the policy instruments help the policy-making part of the system to learn and adapt most effectively. These and other factors determine whether 'subsidies' such as those provided through these instruments make the complex adaptive system more or less resilient.

Finally, it is worth noting that in addition to the effects of the instruments on land management mentioned above, they also reduce the signals encouraging producers to take up new self-reliance technologies. In another study of the value of new seasonal forecasting methods to pastoralists, we found that significant benefits, which could be achieved by using the forecasts in terms of cash flow, almost vanished once these tax instruments were taken into account. Here we see that the current baggage of instruments may significantly inhibit uptake of more 'self-reliant' behaviour— another example of co-evolution (not currently fully adaptive) between elements of the complex system.

Discussion

In the conclusion to our report,[15] we wrote words echoing those I'd written eight years earlier to the Drought Policy Review Task Force, that trying to support the industry when it was *in* drought would always suffer from the need to reconcile the irreconcilable— balancing equity and objectivity in assessing when to give funds.

Consequently, we argued, putting the same funds in to long-term self-reliance and drought-preparedness measures would be a far more effective and less contentious action, although this might need to be coupled with a better strategic assessment of those regions in which it is possible to deal with drought adequately and stay viable. That was not, we were at pains to point out, to suggest that families should be left in personal hardship, but that this aspect should be dealt with through the established social security system rather than in a separate special exceptional circumstances process for agriculture. A decade on, I don't see any reason to change this view, but can back it up a bit more with some studies and theory to justify the position in a broader framework.

Why should we support an industry in extreme conditions at all? We need to be very clear about this—from the point of view of supporting the industry (as opposed to post-productivist goals such as occupying country, sustaining culture, and so on[16]) it is justified when the temporary public cost of supporting it is (substantially) less than the ongoing public benefits in having the industry around. Recent National Land and Water Resources Audit data[17] suggest that many regions of the rangelands have only small positive or even negative total terms of trade as pastoral enterprises when the full costs of production including public inputs are taken into account[18], totalling only $100 million per year or so in net productivity across all rangelands (not the $1 billion figure of gross production usually quoted). This doesn't even include the capital costs of resource degradation, even if there is good evidence that this may be slowing. In this case, spending large amounts of public resources on industry support is not justified. The total Commonwealth cost for agriculture of all the taxation instruments alone is estimated at $250 million per year,[19] with a somewhat disproportionate part spent in rangelands; this doesn't include any special drought or rural restructuring funding, let alone the matching state funds. Of course, there are other reasons for rangelands managers to be on the land, but in that case we should be supporting those reasons, not animal production.[20] Written less harshly, the same argument will apply to other areas of Australian agriculture.

So the first point is that we need to be clear about whether support is warranted at all. If it is, then we must work out what needs supporting—do we provide financial support, the main focus of the past (in forms from direct transaction subsidies to income averaging and distorted stock valuation allowances), or something else? Here we must be wiser in tailoring our support to the features of our landscapes, our people, and the complex adaptive systems that they all constitute.

From the previous case studies and other experiences, there seem to me to be some key guidelines for policy development in this area. These are undoubtedly based on a very partial analysis, with an eye to the long-term rather than the transitional arrangements that might be needed to get there, and are intended to stimulate debate rather than curtail it. They also are biased clearly towards my rangelands experiences, although it must be said that the rangelands have managed always to take the lion's share of drought subsidy, angst and political attention in comparison to their relative productive capacity.

1   *Match the spatial scale of intervention to the scale of the social and ecological processes of concern.* Blanket national policy instruments, such as taxation policy, are very blunt instruments for achieving subtly differentiated changes in ecological management at a property-by-property level. These types of instruments need to be focused, where appropriate at all, on changing behaviour in more universal domains of management, such as financial management. Where this is impossible without affecting ecological management, they should not be attempted at all.

2   *Match the temporal scale of intervention to the social and ecological processes of concern.* The temporal scale for managing drought is at least decadal, since it must be founded on preparations through the full pseudo-cycle of wet and dry times. Instruments that focus on just one part of the cycle (like exceptional circumstances, livestock elections and so on) inevitably fail the test of holism. Additionally, policy-makers need to be able to adjust their instruments adaptively over time as conditions change; this is unlikely to be achieved if the instruments are targeted at factors that change fast, and if policy-makers

do not have the appropriate feedback (or ability to interpret it) on the effectiveness of instruments.

3   *Create an environment for all elements of the system to learn.* There are many places in the modified adaptive cycle where managers, their organisations, the community and their policy-makers at all levels need to be able to understand the consequences of their previous actions and co-evolve their behaviour. A key role of policy should be to create an enabling environment for this learning, in their own experiences as well those of others. Given the difficulty of detecting change in variable environments, any actions, instruments or incentives that cloud the feedback are particularly destructive to learning in the Australian environment.

4   *Keep the vision but remember the need for transition.* Systems don't change overnight; sometimes generations have to come and go; and political cycles career back and forth like frantic meat ants on the time scale of Australian environmental change. Nonetheless, if the problems of transition overwhelm the ability to see the road ahead, there is little future prospect for a new Australian agriculture. New policy must contain the seeds of its own change, as well as ensuring it doesn't inhibit others from evolving.

We have a long way to go to implement these lessons and implement them in a way that is socially tolerable in transition. The above points are illustrative of a notable effort emerging as many people over the past decade have begun to make a real link between the ecology and institutional structures of this great southern land of ours. It is exciting to see policy-makers, economists, journalists, and scientists, whether socially or biophysically oriented, sitting down together to try to work out the new paradigms for dealing with the particular social and physical environment of Australia. We are learning to be Australian, with some universal lessons that need appropriate application here, and other lessons that are peculiar to our environment. These lessons include learning to cope with complexity, local differences and variability, and learning to have a learning environment. 'Drought' is complex and multifaceted, and there will be no single drought policy, but rather a series of objectives (ends) and processes (means). We need to keep a focus on a suitable eventual target for drought policy in this variable continent, and work back from that to worry about routes for transition, rather than being diverted by the political barriers that impede the day-to-day steps towards our antipodean future.

1 For example, White DH 2000.
2 Levin 1998.
3 Walker and Janssen 2002.
4 For example, Stafford Smith and Reynolds 2002.
5 For example, Walker *et al.* 2002.
6 For example, Ostrom 1999.
7 For example, see Stafford Smith and Reynolds 2002.
8 For example, Stafford Smith and Foran 1990.
9 Stafford Smith *et al.* 1997.
10 Stafford Smith and McKeon 1996; Stafford Smith and McKeon 1998.
11 See Table 2 in Stafford Smith and McKeon 1998.
12 For example, Buxton *et al.* 1995—there were 8 such reports.
13 Cross and Stafford Smith 2001.
14 Stafford Smith *et al.* 2001.
15 See White, DH 2000.
16 Holmes 2002.
17 NLWRA 2002.
18 For example, Fargher *et al.* 2002.
19 Treasury 2001.
20 Stafford Smith *et al.* 2000.

Perceptions of drought risk: The farmer, the scientist and the policy economist

Peter Hayman and Peter Cox

Introduction

If risk is defined as 'uncertainty with consequences',[1] dryland farming in Australia is risky. Although any human activity that has to allocate resources for an unknown future is by definition dealing with risk, the extent of seasonal variability in Australia leads to a particularly wide range of uncertain outcomes. This uncertainty due to climate has serious consequences for farm businesses, farm families and rural communities. The consequences of the variability have increased due to declining financial margins and higher levels of farm debt. Climate is not only a source of risk to farm businesses, it is a major source of risk to physical resources and social resources. In physical terms, environmental degradation can result if risk is not managed responsibly. Socially, rural communities can become less viable and resiliency can be undermined if farms go under.

The consequences of how drought risk is managed are profound for farm businesses, farm families, rural communities and the landscape. The primary responsibility for managing drought risk lies with individual farmers. Hence, the design and modification of effective policy instruments to improve the management of drought risk in Australia depends in part on understanding how farmers perceive and manage the risks associated with periodic drought. In this chapter we examine these risks, how farmers perceive them, how scientists try to quantify them, how quantification can be used to improve risk management, and some of the pitfalls in relying too much on 'scientific' (objective) models for risk assessment. Mismatches are noted between different perspectives of drought as risk. These mismatches suggest that the issue of drought is still incompletely specified, that its complexity will always allow multiple interpretations, and that useful policy intervention has to respond to these differences and build on them.

Within the framework of the National Drought Policy, appropriate ways to manage climate risk in Australia will differ from region to region. Even within regions, different ways of managing climate risk are inevitable because of differences in farmers' goals and perceptions of risk. Agricultural extension has been criticised for ignoring the diversity in approaches to farming even among neighbours with similar resources,[2] the same critique may apply to policy. Not only is diversity inevitable, it should be encouraged as a means of building resilient farms and communities. Diversity is part of managing risk.

Australian agriculture is a risky business

Across the world, agriculture can be interpreted as an activity designed in part to reduce the risks of hunting and gathering.[3] As the first European settlers struggled to develop farming systems in this strange land with poorly defined seasons, they must have been relieved initially not to have to deal with the winter freeze. The challenge that rapidly emerged was the non-seasonal year-to-year variability, much of it due to the ENSO cycle. This difference is evident in the regular breeding cycle of animals and flowering of plants in the northern hemisphere compared with the erratic opportunistic reproduction in Australian fauna and flora.[4] One of the clearest examples of the difference for humans dealing with this climatic uncertainty is the mid-winter feast which was taken over by the early Christians to become Christmas. In the middle of winter, a lot of food is consumed; the notion of a mid-drought feast would be ludicrous. In the variable climate of Australia, agriculture may be more risky than nomadic hunting and gathering.[5]

After surviving the 1990s, most farmers in Australia would be surprised that there was much to be written about their perception of climate risk beyond stating the obvious. Coping with droughts and floods is likely to be high on the list of advice offered by experienced farmers to newcomers. As outlined in Chapter 5, droughts are milestones in the lives of rural people. Rural people will often talk about events prior to or after 'the drought' of the 1980s or 1990s. In eastern Australia, years such as 1994, 1982, 1977 and 1965 spark immediate recognition. These events are remembered because drought, unlike bushfires and floods, is a relentless slowly developing phenomenon that results in a series of stressful decisions, business failures and neighbours (even partners) leaving never to return.

Drought is a recurring theme in Australian history[6] and a feature that distinguished Australia from England. In Britain in 1887, a drought was defined as 15 consecutive days, each with less than one point of rain (0.25mm).[7] The NSW Government Astronomer, H.C. Russell, drew attention to a colonial Australian concept of drought:

> *The word drought is not used here as in the sense in which it is often used in England and elsewhere; that is, signifying a period of a few days or weeks in which not a drop of rain falls, but it is used to signify a period of months or years when the country gets burnt up, grass and water disappear, crops become worthless and sheep and cattle die.*[8]

The emphasis on the consequences of rainfall deficiency, not just the low rainfall, is a feature of how Australians view drought and makes the assessment of drought risk, and the development and communication of coherent policy, a challenge. The concept of drought risk in Australia would be simpler (although not necessarily better) if the understanding of drought was restricted to rainfall as in the British definition.

One of the important developments in the understanding of drought in Australia was the National Drought Policy (NDP) of

1992, and the deliberations of the taskforce leading up to it. This policy was the first to place emphasis on farmers' risk management (see Chapter 3 in this volume). This entailed viewing drought and fluctuating seasonal conditions as a normal recurring phenomenon that should be planned for just as any other business risk. It attempted to change the perception of risk away from the government as an insurer of last resort[9] towards seeing risk as part of the environment within which production decisions have to be made. The role of government was switched from one of disaster relief during a drought to one of ensuring that farmers are equipped to manage climatic risk effectively themselves, with the safety net of support in times of 'exceptional circumstances' for which farmers cannot be expected to prepare.

Research and extension providers were funded to assist this transfer of the responsibility for managing climatic risk from government to farmers. One of the first activities following the NDP was a workshop on incorporating risk assessment and analysis into decision support systems and farm business management.[10] Further progress on risk management and communication was made at a Workshop on Drought and Decision Support held in Canberra[11] and the Risk Management in Agriculture Conference at Armidale.[12] The role of publicly funded research and development in providing information (rainfall probabilities, simulated crop and pasture production data based on historical climate records) and training (workshops, decision support systems) was generally treated with enthusiasm. One exception was the farm management economist, Bill Malcolm,[13] who maintained that there were two untested assumptions underpinning much of the research and development in the wake of the NDP: that farmers were poor managers of risk; and that R&D could help them manage risk better. He pointed out that, even if the first claim were true, the second did not necessarily follow. The NDP encouraged communication between farmers and researchers about the risk of drought and the riskiness of different decisions for managing it. But this was not as straightforward as it first appeared.

Different ways of defining risk

It is not surprising that economists, scientists and farmers struggle to find a common language for concepts of risk management, especially on an issue as emotive as drought. Just as farming (like science and economics) is socially constructed, the riskiness of farming is a negotiated construct that cannot be understood independent of our minds and culture. What is possible, and what is considered appropriate behaviour, depends on who is looking and the group they belong to. Furthermore, risk as a term can be used in both a technical-legal sense and in everyday language. Although different meanings of risk are a challenge for communication and policy direction, acknowledging the differences can enrich the process. The danger to communication is when differences are ignored. Risk is broadly defined here as uncertainty with consequences. Risk management involves reducing uncertainty, coping with variability, avoiding peril and exploiting opportunities.[14]

Aspects of risk that impact on policy development include: defining risk as chance of loss *versus* risk as variance; a distinction between risk and uncertainty; treating the sources of risk in isolation or as multiple interacting causes; the representation of risk as a probability distribution that provides a description of the world external to the decision-maker or as a psychological process; and drought as a local production constraint or a set of spatially correlated impacts.

Risk as chance of loss versus risk as variance

The Macquarie dictionary defines risk as chance of loss[15] and this is probably the general use of the term. Among economists, risk refers to the variability (variance) in the outcome that results from an action. Risk analysis considers a trade-off between the mean of the outcome and its variance or, in the case of stochastic dominance, between the patterns of distribution of different outcomes. A distinction between risk (variance) and down-side risk (negative deviation) is sometimes made in the economic literature.[16] However, if the distribution of outcomes is symmetrical (for example, a normal distribution), it makes no difference and parsimony would suggest that the distinction is not required. If distributions are highly skewed, or if some outcomes are catastrophic, use of more complex analyses based on concepts such as negative semi-variance or negative absolute deviation may be appropriate. A symmetrical distribution of crop yield is likely to lead to a skewed distribution of profits, which will change again when tax is considered. The shape of the distribution of outcomes will not matter to risk-neutral decision-makers who, by definition, are only interested in maximising mean wealth and are unconcerned about the variability/risk involved. Needless to say, most people are risk averse; that is, we care about the down-side risk as well as long-term averages of outcomes. Economic studies suggest most Australian farmers are slightly risk averse.[17]

Risk comes from the Italian *risicare*, which means to dare and emphasises choice, opportunity and gain rather than fate and loss.[18] In general, variability is not a bad thing: it sometimes allows us to recover from our mistakes; it can be a source of novelty; and it provides a screen against which to choose between alternative decisions. Furthermore, the ability to handle variability is one of the sources for private entrepreneurs to gain competitive advantage and succeed.[19] The emphasis on risk management in drought policy sees drought as something that farmers can win from. The bad years are the cost of the good years. But the bad years can be managed so that the returns in the good years are even greater.

Risk versus uncertainty

Another aspect of risk that has policy implications, especially for the funded research and development in the wake of the drought policy, is the distinction between risk and uncertainty introduced in 1921 by Knight.[20] A simple example is considering the toss of a fair coin as risk, whereas the toss of a biased coin is uncertainty. After experimenting with the biased coin, the uncertainty could be quantified as risk.

This notion underpins much of the role of science assumed in the NDP. The reasoning is that the future is uncertain, and that, using rainfall records and crop and pasture models, some of this uncertainty can be quantified as risk. Central to this approach is the frequentist view of probability distributions, which uses historical data (measured or simulated) to produce a probability distribution of outcomes. The alternative, subjectivist view uses probabilities to capture the degree of belief an individual has that a given outcome will occur.[21] The subjectivist view is not only expedient, but a legitimate way to construct a decision calculus. Decision-makers may adopt the output from a scientific model or historical rainfall data, but this then becomes their subjective view.

The collation or modelling of derivatives of rainfall, such as animal or crop production, requires considerable judgement by scientists and trust in that judgement by farmers and advisers. These judgements are usually not quantified and, in a frequentist sense, hardly quantifiable.[22] The trust requires an assumption that historical records are still applicable and that the models used reliably transform these data into probabilities of consequences. The contribution of biological scientists is partly based on the idea that it is both possible and desirable to specify objective probability distributions that describe critical system parameters. However, for many farm management problems, this is only partly possible (because the system can only be partially specified in an engineering sense), and may not even be the most effective means of intervening and improving the management of farming systems.

Treating isolated sources of risk or multiple interacting sources

Although textbook treatments of decision-making under uncertainty often deal with a single source of risk, which has a fixed relationship with the outcome (for example, the impact of seasonal rainfall on farm returns), this is rarely the case in practice. Climatic variability is a major contributor to agricultural production risk, but other factors also influence production risk, such as pests and diseases or responsiveness to fertilisers. From the point of view of farm management, production risk is only important if it affects business risk. The other important component of business risk is price risk. Historically, this has been the major focus of government intervention and was rated in a 1993 survey of farmers across Australia as the major risk of concern to farmers.[23] A further complexity is added by consideration of financial risk, which is the variability of net returns to owners' equity after financing debt. During the drought of the 1990s, interest rates for some farmers were as high as 23 per cent per annum. One experienced financial consultant included in a more holistic risk assessment the risk to family labour 'who are underpaid and in many cases overworked and ill rewarded for all their efforts, but trapped into trying to preserve the family capital through desperate measures'.[24] Others have added the considerable occupational and safety risks of rural work and the risk of divorce. None of these risks is rare, but they are exacerbated by the stress of drought, and are often fatal for the farm business.[25]

Consistent with a more holistic view of risk, the adjustments to exceptional circumstances (see Chapter 3) led to tighter mathematical definitions of what was meant by rare events, but broader definitions of factors that could be considered as exceptional circumstances. This is supported by whole-farm stochastic modelling that showed drought as just one factor causing serious decline in net farm income.[26] However, it has led to some angst amongst farming communities, as more judgement is required, and hence the process is perceived to be more readily politicised and bureaucratic.[27]

Alternative and sophisticated holistic thinking on drought risk comes from work in developing countries where the collision between drought and poverty is dramatic. One of these approaches, vulnerability analysis, turns conventional impact analysis on its head by considering multiple causes of critical outcomes—dislocation, hunger, famine—rather than the multiple outcomes of a single event such as drought.[28] Another distinction is made between risk as *ex ante* income management and coping with bad outcomes through *ex post* consumption management.[29]

In an Australian context, Malcolm argued for a broader view of risk and suggested that the recent enthusiasm for risk management was partly due to misinterpreting problems of low farm income as being due to poor risk management. Poor income is more likely to be due to structural problems at farm and industry levels. While climatic risk may exacerbate the problem, information and procedures on climate risk will do little to solve the problem if the underlying cause of low income is farm size or inappropriate land use.[30]

Representing risk as a probability distribution or a psychological process.

Much recent research into the psychological process of risk perception in agriculture relates to quarantine,[31] biotechnology[32] and pollution.[33] A psychometric model of risk uses scaling techniques to systematically measure responses to a series of hazards. These methods have shown that hazards judged as dreadful and unknown are also judged as the most risky.[34] Psychological studies have identified various issues that influence the perception of risk, including the subject's sense of control and her worldview, whether a risk is voluntary, and the distribution of costs and benefits. Feelings about, or response to, risks are central to a lifetime of learning:[35] the point about learning is finding out progressively what is and what is not possible, and using this knowledge to change behaviour; that is, manage risk.

Drought as a local production constraint or a set of spatially correlated impacts

Drought risk is not an extreme irreversible event like global warming and the key management decisions (de-stocking or modifying crop area) are at a local level rather than requiring intervention at a regional or national scale (compared with the introduction of an exotic pest that can spread rapidly through a wide area). Nevertheless, the effects of drought are usually correlated spatially. This is used as an argument for political intervention. Because many people are affected by drought simultaneously, it is a matter of public concern. However, it may in part be an indication that there is not sufficient diversity in the way people manage risk. Drought is only a source of risk because of the way it interacts with artificially made systems such as agriculture and the way it impinges on the way people see, and act in, the world.

Different perceptions of drought as a source of risk

So far in this chapter we have established that risk is a negotiated construct. If drought policy is based on influencing how individual farmers manage risk, we need to understand underlying cultural perspectives that colour the way in which drought is perceived as a risk and hence how the changed signals will be interpreted.

Drought as a risk to food security by pioneering farmers battling harsh elements

An analysis of hundreds of Australian media articles, parliamentary speeches, books, poetry and films suggested that drought was regularly invoked as a symbolic threat to the Australian national community.[36] Furthermore, this symbolic use of drought showed no sign of waning, despite the declining relative importance of agriculture to the national economy. In another analysis of the wording of newspaper reports of droughts from 1900 to 1995 little had changed.[37] Droughts are declared; drought is something we must combat and battle with a plan of attack. Drought grips, creeps, bites and decimates the land and people who are drought-smitten, desperate, ruined. Land that is irrigated or in higher rainfall areas is called safe country—safe from the 'dread enemy', as drought was called in 1906, or 'our biggest enemy' in a headline in 1995. If drought is a similar threat to war, the nation is at risk and government intervention is easily justified. In the context of the 1994 drought, a senator described drought as a time for all levels of government and all sides of politics to work together through a national tragedy.[38] In December 2002 John Ubergang, a farmer from northern NSW, proposed the Crooble Plan[39] to avoid the 'major social and economic chaos and disaster for Australian rural producers and small businesses'. He proposed that the Federal government should take 3 percent of the national budget to fund a National Disaster Management Program, which would be allocated by a team of local farmers, graziers, agronomists and veterinary surgeons once any local weather station recorded a two-monthly rainfall below 50 percent of the long-term average for those two months.

Drought as a risk to the efficiency of the rural, and hence national, economy

Agricultural economists[40] argue that (a) the underlying assumptions of risk to the breeding stock and skilled labour through a drought are generally overstated; (b) drought assistance and transport subsidies can lead to environmental risk by overstocking and keeping stock too long into a drought; and (c) assistance leads to moral hazard; that is, a cross subsidisation of the careless by the careful and, in some cases, of the dishonest by the honest. If climate is treated as a unique source of risk requiring assistance, it is likely that there will be too much investment in farming drought-prone regions, compared to regions with more reliable rainfall. Furthermore, across all regions, farmers are likely to pay too little attention to managing climate risk compared to other risks if this source of risk has already been hedged by government policy.

Nevertheless, the underlying notion of keeping the farm going for efficiency rather than welfare reasons is strongly held. A recent example is the Queensland Western Downs Solutions Group formed in March 2002. In their request for a one-off injection of $10 million and reforms of drought assistance, the justification for funds was based on the 'past and future contributions of the drought affected shires to the economy'.[41] In this sense, drought is a risk to the economy and drought support is part equity for past contributions and part investment for future contributions. Efficiency and equity in terms of the adjustment process was used by the NSW Farmers' Association in their submission to a review of assistance arrangements:

> *The well-acknowledged rationale for public assistance in this case is that it prevents inefficient adjustment that might otherwise occur. Australian farmers receive very little Government assistance, especially compared to farmers overseas. It is therefore appropriate that they receive assistance to withstand events which may cause them to exit the industry, if they are otherwise viable.*[42]

Drought as a risk to the welfare of rural families and communities

Drought is recognised as a factor in divorce, suicide and illness in rural areas.[43] Drought serves as a catalyst for major upheavals in rural communities; it is the focal point for structural problems of farm size, the cost–price squeeze and the fragile interdependence of rural communities. To those feeling these compounding effects, drought is an intense lived experience, rather than part of a probability distribution or risk profile. In the context of researching lived experience, Virginia Woolf's description of a metaphor has been used as a means of not describing the object itself but providing 'the reverberation and reflection … close enough to the original to illustrate it, remote enough to heighten, enlarge and make splendid'.[44] In this sense, drought is a metaphor for rural hardship and suffering.

When then Prime Minister Paul Keating pointed out in 1994 that drought and climatic variability are part of the natural environment and did not constitute a natural disaster, he was following a line that demystified drought. At the time, he was criticised by the media for a callous comment. An indignant opposition senator claimed he was 'condemned all over Australia by every person in the rural community for saying that drought is a natural recurring phenomenon'.[45] Although this was partly party politics, when a policy economist asks a farmer to consider drought as a normal recurring business risk, some farmers interpret this as being asked to take the enlarged metaphor of rural suffering as a recurring risk.

Drought as a risk with a unique place in rural society

Each culture accepts some risks as normal and views others as catastrophic.[46] Over most of history, climate was understood to be controlled supernaturally and extreme events taken as a sign of God's wrath. The number of farmers who have attended prayer meetings for rain probably exceed attendance at scientific workshops on El Niño. Indeed, El Niño is regularly personified in cartoons and press stories. Jill Ker Conway[47], in her autobiographical account of the drought in 1940s, in which her family lost the farm and her father died probably as a result of suicide, observed that the drought caused her 'to lose her faith in a benign providence' and as a young girl to take to heart Shakespeare's King Lear: 'As flies to wanton boys are we to the gods. They kill us for their sport.'[48] This sentiment resonates with Henry Lawson's poem Beaten Back:

Can it be my reasons rocking,
for I feel a burning hate
For the God, who only mocking –
sent the prayed for rains too late.[49]

Other sources of uncertainty, such as price variability, may have a similar impact, but do not evoke the same depth of emotions as climatic variability. This contrasts with the neoclassical economic treatment of climate variability in which a perfectly informed society adopts optimal strategies.

Drought as a risk to animal welfare and the land

Drought is an extremely high-risk decision environment, especially for graziers. As the drought worsens, prices fall and the cost of feeding increases. The decision to sell is made more difficult by concern that prices might rise when widespread rain comes. The advice in the past could be summarised as either 'sell and regret, or let the worthless buggers die'.[50] The second option is now illegal on animal welfare considerations. In any case, the classic drought photo of the skull on parched ground represents a situation that should not occur, as stock should be removed long before they or the groundcover dies.

Most rural producers are concerned about the environmental impact of drought. However, it is likely that more and more urban taxpayers are likely to see drought as a risk imposed by people on the environment. There is an increased preparedness to challenge current farming practices (a situation not dissimilar to the public reaction in Britain to BSE and foot and mouth disease where sympathy for farmers was mixed with a challenge to farming practices). Following an article in the *Sydney Morning Herald* on the drought and the request from farmers for assistance in July 2002, two letters appeared, neither of them sympathetic. The first letter[51] observed:

> *Very little time seems to pass between claims that farmers are in trouble and having a hard time. How often have we seen them receive financial assistance because of a drought, flood, loss of stock or crops? Compare that with the number of times we hear about small business being offered money to enable them to continue trading … If the cockies find life so hard on the land, why don't they leave it and try to make a life in the city, as many of us are trying to achieve?*

The second letter[52] expressed concern over the environment:

> *Australia is periodically affected by drought due to the cyclical El Niño climatic effect. This has been going on for tens of thousands of years. However, Europeans have been settled here for only 214 years and appear not to have adjusted to this phenomenon … One could ask when are we going to put limits on intensive agricultural development in this fragile, drought-prone country?*

The drought of 2002 has been labelled the first green drought[53] in reference to discussion on appropriate ways of managing Australia's land and water resources. This debate became prominent when leading businessmen and media personalities launched the Farmhand appeal for drought-affected farmers and proposed ways of drought-proofing Australia. Media coverage and the debate was maintained by the response by the Wentworth Group of Concerned Scientists. It is likely that urban taxpayers will increasingly join the debate on drought risk between farmers and governments.

Different assessments of the likelihood of drought

Although there are many facets to risk, most approaches to drought risk have some notion of the likelihood of low rainfall. A view of drought as a mismatch between farmers' expectations of rainfall and the rainfall that occurs implies that farmers tend to be over-optimistic and misjudge the variability of the climate. Supply of objective rainfall probability data by researchers is held to correct this misperception. The argument for Australian farmers misjudging the variable climate comes from the muddle of past drought declarations, historical analysis of a pioneering optimistic sprit, literature from the field of psychology and limited empirical evidence.

The muddle of past drought policies

The history of official drought declarations suggests the refusal to accept the reality of the arid and variable climate—a reality that it is in abnormal, rather than normal, seasons that Australia is green like England. In Queensland, some shires have been either partially or completely drought declared 70 per cent of the time from 1964 to the early 1990s. In NSW, some districts have been drought-declared for three months or more for 65 per cent of the time.[54] There is a striking difference between these frequencies of drought declarations and the meteorological definition of drought as the driest 10 per cent of years in the historical record,[55] or the definition of exceptional circumstances as the driest 5 per cent of rainfall for a defined period. Some care must be taken in using drought declaration for evidence of misjudging climate, as it might be just effective advocacy. Recalling his days as Finance Minister in the 1980s, Peter Walsh[56] referred to drought declaration as a regional licence to milk the Canberra cash cow. Some of the inconsistencies between states have been analysed[57] and support Walsh's assertion.

The Goyder line as a warning to agricultural optimism

The 'Goyder line' was named after the first Surveyor General of South Australia who, in 1865, established a line to mark the limit that drought had extended south. This was done to assist in determining which pastoral leases might be given financial relief.[58] It became a line beyond which cropping should not proceed. It was ignored during a run of good seasons with dire consequences. Governor Phillip's response to the first drought he experienced was that he did not think it probable that so dry a season often occurred. This optimistic view of climate has characterised much of the pioneering spirit in Australia.[59]

Evidence from psychology that humans find it hard to think about risk and uncertainty

People (not just farmers) are poor intuitive statisticians. And most of us refuse to admit it. Shanteau[60] noted the widespread evidence (including studies with farmers) that people have limited cognitive capacity to process low probabilities and focus more on the magnitude of single outcomes than the likelihood of different outcomes (see Table 8.1). These biases have been related to people's understanding of climate variability.[61]

Table 8.1 Perceptual biases[62]

Bias	Description
Availability	Availability bias means that items or events presented first (primacy) or last (recency) assume undue importance. The proximity of a drought or a run of good seasons is likely to affect the rated frequency.
Selective perception	People seek information consistent with their own views. Many farmers have strong views on cycles of floods and droughts, and look for confirming evidence. Scientists look for confirming evidence of the value of seasonal climate forecasts.
Concrete information	Vivid, direct experiences dominate abstract information; a single personal experience can outweigh more valid statistical information. Farmers are likely to remember the booms and busts more than the average years.
Law of small numbers	Small samples that are readily available to a decision-maker are seen as representative of the larger population even when they are not.
Insensitivity to base rates	In dealing with uncertainty, people often ignore background information; for example, that there are about twice as many El Niño events as there are bad droughts in a given location in eastern Australia.
Gambler's fallacy	People can be convinced of patterns that don't exist; for example, that random events are self-correcting. By the autumn of 1994, farmers in southern Queensland had had three bad seasons; when it rained in autumn they believed that they were due a good season only to get the worst El Niño drought since 1982.

Most of these biases relate to imperfect sampling, whereby we are prone to use the wrong distribution to derive the likelihood of an event. On this basis, science with easy access to long-term climate records appears to have the upper hand on farmers who have unreliable memories of limited time series. However, this argument must be used with caution. Due to climate change the underlying sample may be changing, there are clear shifts in temperature, evaporation and, to a lesser extent, rainfall.[63] Furthermore, the sample of wheat yields or carrying capacity may not be all that simple and straightforward due to technology trends and complex interactions with degradation events. Some of these changes can be captured in simulation models but others cannot. A counter argument to the gambler's fallacy for climate is the notion that the oceans, the engines of periodic drought (unlike a coin) do have memory. In any case, some agricultural processes are self-correcting, such as putting on fertiliser that is not used for a crop but is available for the following season.

In a previous text on drought policy in 1973, Heathcote[64] hypothesised that 'folk' perceptions of drought risk are much lower than the 'scientific' assessment of drought risk. There is some support from a study in Western Australia in 1984, which found that drought frequency was underestimated and viewed incorrectly as having a regular cycle rather than being episodic.[65]

A different finding to the farmer as a misguided optimist emerged from interviews with 90 farmers and 20 advisers in northern NSW, to document their subjective risk assessment of seasonal rainfall and derivatives of rainfall such as fallow recharge and crop yields.[66] When these subjective risk assessments were compared to the long-term rainfall record and simulated yields and fallow recharge, farmers saw the climate as drier and more risky than the long-term records reflect; they rated the chance of crop failure and low yields much higher than crop simulation models did. This contrasts with conventional wisdom, which holds that farmers are overly optimistic and need access to the long-term rainfall record to appreciate the true risks of farming. Furthermore, most farmers interviewed had lived through a wetting trend. A farmer with perfect memory of only the last 20 or 30 years might be expected to view the climate as wetter than the 100-year record suggests. Such evidence, as we do have, shows that farmers may be too conservative in the allowance they make for the possibility of drought. In any case, farmers may be able to adapt quickly to the different set of outcomes engendered under the NDP, with or without access to additional quantification or transformation of rainfall data.

Different models of managing risk

From a psychological perspective, Hammond[67] argued that uncertainty in the world outside the observer generates uncertainty in the observer's cognitive system. This view allows for a continuum in the uncertainty of the external environment from a highly controlled environment (for example, a chemical engineering plant) to a natural environment. As the discipline of agricultural science shifts from reductionist science to the study of farming systems and then to applied ecology, the quantification of uncertainty as risk becomes more difficult. Systems are less well specified; there are far more linkages, including feedback and feed-forward loops; and we are dealing with competing claims about what will happen and what should happen. Hammond's systems view of risk argues that cognitive systems (the way we order experience and construct reality) also have a continuum from intuition, which takes little account of uncertainty, to analytical thought, which in extreme forms such as decision analysis attempts to account formally for uncertainty.

There is a mismatch between the understanding of risk by farmers dealing with complex systems with clever but relatively simple intuition,[68] and the treatment of risk by scientists modelling relatively constrained agricultural production systems with complex formal analysis. In terms of managing risk for a farm business it is better to solve the whole problem roughly than attempt to solve part of the problem extremely well. Even at the whole-farm level, risk is only a partial issue. Farm management economists have argued that research and development may have swung from ignoring risk to placing too much emphasis on risk.[69]

If what we are saying is correct, the marginal returns to more elaborate formal analyses to improve the efficiency of drought management may be modest. The key issues may be elsewhere— finding the best way to encourage farmers and their communities to make their own decisions, but provide a welfare safety net; determining how environmental and production values can be managed together during a drought; a switch from analysing micro trade-offs between risk and return within a given enterprise (for example, fertiliser rates or crop choice), to encouraging a diversity of responses to drought, including diversification of rural livelihoods;[70] or deciding the extent that an urban population values farming communities and their contribution to natural resource management.

Conclusions

This book has three broad themes. In closing, we relate the issues of perceptions of risk to these themes.

Learning to be Australian

Australian farmers are men and women of action. Farmers with whom we discussed this chapter tended to dismiss worrying too much about words on drought risk. However, a linguist[71] studying drought noted that our language is set to a default country, which is narrow, green, hilly and wet, where rivers run, lakes are full and drought is an exception. She maintained that if we use the language of war and invading pirates for a natural recurring phenomenon like drought, 'Australia is a place perennially disrupted, disjointed and the experience of Australia must always be one of disappointment and suffering'. Heathcote[72] noted the indignant surprise with which each drought was treated and suggested the cause was part psychological, part patriotic. Daly[73] argued that the response to drought indicates Australian society can only deal with the good times. There are many rural Australians who have developed a mental picture of agriculture that fits all facets of the risky climate. Although not easy, this is necessary for their psychological health and genuine empathy with the country they live in.

While it may be true that Australia has a more variable climate than Europe and much of northern America, it is problematic to treat climatic risk as a problem unique to Australia. Although there may be some comfort in a narrative that holds Australia as special, this may be isolationist. A simplistic approach to being Australian ignores the experiences of many poor countries in South America, India, Africa and parts of Asia that face similar inter-annual variability in rainfall, but are far more vulnerable to climatic risk. Learning to be Australian involves understanding the ways we are different and similar to other communities.

Drought is complex and multifaceted

If drought is complex, it is unlikely that we can capture it simply in a single probability distribution or risk profile. The US Environmental Protection Authority maintained that a hard lesson for technical experts was to acknowledge that, in the process of risk characterisation, apparently inexplicable inconsistencies may be recognised as responsible, reasonable descriptions of the same problem.[74] When scientists use terms such as 'real risk' and 'perceived risk' to dismiss people's concerns about pesticide use, they have been accused of 'unreflective use of ambiguous, emotion-laden words from everyday language'.[75] That is not to say that the contribution of science to assessing and managing drought risk is trivial; rather, it is partial and will be most useful when it is clearly stated as partial. In recognising that it is partial, we must acknowledge that the management of climatic risk may not always be limited by information.

There is no single drought policy, rather there is a series of objectives (ends) and processes (means)

Have farmers' perceptions changed? Improving farmers' risk management was a process towards the goal of self-reliance. In a booklet entitled *Taking control*, funded in the wake of the NDP through the Property Management Planning Campaign, Richard Golden, a farmer who joined the program at the end of the drought of the 1990s in southern Queensland claimed, 'I have just about wiped out the victim mentality which I believe still besets much of agriculture'.[76] The campaign reached about 13 per cent of farmers. When surveyed, 78 per cent of farmers in SA and 87 per cent in NSW indicated that they were more self-reliant as a result.[77]

The process of increased engagement between rural producers and the professional R&D establishment following the NDP did encourage scientists to consider risk in their analyses and to develop better ways of engaging with farmers.[78] We have argued that the exact specification of probability distributions may be a more modest contribution than first thought. However, probability distributions of rainfall and production have started to provide a common language and a Rosetta stone for communication about risky decisions. For the biological researcher, it opened up the uncertain decision-making world of practical farming, and farmers were introduced to more methodical ways of framing farm management decisions. It helped clarify the thinking of farmers, researchers and policy-makers about our environment and issues of resource allocation. It helped challenge long-held assumptions about the nature and importance of drought, risk and the role of publicly funded R&D.

If part of the policy response to drought is R&D funding to improve farmers' management of farming systems in a variable climate, explicit treatment of risk as uncertainty of outcome for management decisions is essential. While a probability distribution on a computer screen may be a contribution, we must recognise that risk is, and will always remain, a negotiated construct that cannot be measured outside of the mind and culture of the decision-maker.

1 Anderson and Dillon 1971.

2 Vanclay *et al.* 1998.

3 Hardaker *et al.* 1997.

4 Nicholls and Sellers 1991; Flannery 1994b.

5 Flannery 1994, p. xx.

6 Nicholls 1997b.

7 Heathcote 1973.

8 Foley 1957.

9 Moss 2002.

10 Trapnell and Fisher 1993.

11 Bryceson and White 1992.

12 Powell 1994.

13 Malcolm 1992; Malcolm 1994.

14 Clark and Brinkley 2001.

15 Delbridge *et al.* 1991.

16 Hardaker *et al.* 1997.

17 Bond and Wonder 1980; Bardsley and Harris 1987.

18 Bernstein 1996.

19 Malcolm 1994.

20 Knight 1921.

21 Hardaker *et al.* 1997.

22 Matthews 2000.

23 Scott 1994.

24 Peart 1992.

25 Anderson 1994.

26 Thompson and Powell 1998.

27 NSW Farmers Association 2001.

28 Ribot 1996; Blaikie *et al.* 1994.

29 Webb *et al.* 1992.

30 Malcolm 1994.

31 Finucane 2000.

32 Coakes and Fisher 2001.

33 Hattis 1994.

34 Finucane 2000.

35 Damasio 1994.
36 West and Smith 1996.
37 Arthur 1999.
38 Brownhill 1994, p. 1681.
39 Ubergang 2002.
40 Freebairn 1983; Simmons 1993. Freebairn 2002.
41 Australian Broadcasting Corporation 2002.
42 NSW Farmers' Association 2001.
43 Munro and Lembit 1997.
44 Van Manen 1990.
45 Senator D Brownhill cited in West and Smith 1996.
46 Douglas and Wildavsky 1982.
47 Ker Conway 1989.
48 Shakespeare 1996, Scene 1, Lines 36–7.
49 Lawson 1988, p. 52.
50 Anderson 1994.
51 Cohen 2002.
52 Bensen 2002.
53 Megalogenis and Wahlquist 2002.
54 Simmons 1993.
55 Gibbs and Maher 1957.
56 Walsh 1994.
57 Smith *et al.* 1992.
58 Andrews 1966.
59 Nicholls 1997b.
60 Shanteau 1992.
61 Nicholls 1999; White 2000.
62 Adapted from McCall and Kaplan 1990; Nicholls 1999.
63 Hennessy *et al.* 1999.
64 Heathcote 1973.
65 Conacher and Conacher 1995.
66 Hayman 2001.
67 Hammond 1996.
68 Cox 1996; Gigerenzer and Todd 1999; Cox *et al.* 1995.
69 Pannell *et al.* 2000.
70 Ellis 2000.
71 Arthur 1999.
72 Heathcote 1973.
73 Daly 1994.
74 Patton 1993.
75 Beck 1992.
76 Grimson 1999.
77 Cock 2001.
78 Nelson *et al.* 2002; McCown *et al.* 1998.

9

Drought policy and preparedness: The Australian experience in an international context

Donald A. Wilhite

Drought is a normal part of the climate for virtually all climatic regimes. It is a complex, slow-onset phenomenon that affects more people than any other natural hazard, and results in serious economic, social and environmental impacts. Drought affects both developing and developed countries, but in substantially different ways.[1] The impacts of drought are often an indicator of non-sustainable land and water management practices, and drought assistance or relief provided by governments and donors can encourage land managers and others to continue these practices. This often results in a greater dependence on government and a decline in self-reliance.

Many people consider drought to be largely a natural or physical event. In reality, drought, like other natural hazards, has both a natural and a social component. The risk associated with drought for any region is a product of both the region's exposure to the event and the vulnerability of society to the event. Exposure to drought varies regionally and there is little, if anything, we can do to alter its occurrence. The natural event, commonly referred to as meteorological drought, is a result of the occurrence of persistent large-scale disruptions in the global circulation pattern of the atmosphere that result in significant regional deficiencies of precipitation over an extended period of time.

As vulnerability to drought has increased globally, greater attention has been directed to reducing risks associated with its occurrence through the introduction of planning to improve operational capabilities (that is, climate and water supply monitoring and building institutional capacity) and mitigation measures that are aimed at reducing drought impacts. Typically, when a natural hazard event and resultant disaster has occurred, governments and donors have followed with impact assessment, response, recovery, and reconstruction activities to return the region or locality to a pre-disaster state. Historically, little attention has been given to preparedness, mitigation, and prediction/early warning actions (that is, risk management) that could reduce future impacts and lessen the need for government intervention in the future. Because of this emphasis on crisis management, many societies have generally moved from one disaster to another with little, if any, reduction in risk. In addition, in drought-prone regions, another drought event is likely to occur before the region fully recovers from the previous event.

Vulnerability is determined by social factors. Population is not only increasing, but also expanding from humid, water surplus climates, to more arid, water-deficient climates and from rural to urban settings for many locations. Urbanisation is placing more pressure on limited water supplies and the capacity of water supply systems to deliver that water to users, especially during periods of peak demand. An increasingly urbanised population is also increasing conflict between agricultural and urban water users, a trend that will only be exacerbated in the future. Increasingly sophisticated technology decreases our vulnerability to drought in some instances, while increasing it in others. Greater awareness of our environment and the need to preserve and restore environmental quality is placing greater pressure on all of us to be better stewards of natural and biological resources. All of these factors emphasise that our vulnerability to drought is continually changing and must be re-evaluated periodically so that we understand how these changes will affect us and who is most at risk for future drought events. As population increases, so does pressure on natural resources. We should expect the impacts of drought in the future to be different, more complex, and more significant for some economic sectors, population groups and regions. Improving drought management implies an attempt to use natural resources in a more sustainable manner. This will require a partnership between individuals and government.

In this chapter I will concentrate on three principal areas. First, there is a discussion of international progress in drought planning and preparedness in the international arena. This will be followed by three case studies—the United States, sub-Saharan Africa and Australia. The latter will necessarily be brief in light of the earlier chapters in this book. I will conclude with some observations about progress in implementing drought preparedness and risk management approaches, including current attempts to establish a global network aimed at improving levels of drought preparedness within and between regions.

Drought policy and preparedness: Overview

Although there has been considerable discussion regarding the adoption of risk-based drought policies and preparedness plans globally, Australia is one of the few countries that have actually implemented national programs or strategies. There are four key components in an effective drought risk reduction strategy.[2] These are the availability of timely and reliable information on which to base decisions; policies and institutional arrangements that encourage assessment, communication and application of that information; a suite of appropriate risk management measures for decision-makers; and actions by decision-makers that are effective and consistent.

Article 10 of the UN Convention to Combat Desertification (UNCCD) states that national action programs should be established to 'identify the factors contributing to desertification and practical measures necessary to combat desertification and mitigate the effects of drought'.[3] In the past 10 years there has been considerable recognition by governments of the need to develop drought preparedness plans and policies to reduce the impacts of drought. Unfortunately, progress in drought preparedness during the last decade has been slow because many nations lack the institutional capacity and human and financial resources necessary to develop comprehensive drought plans and policies. Recent commitments by governments and international organisations, combined with new drought-monitoring technologies and planning and mitigation methodologies, are a cause, however, for optimism. The challenge is the implementation of these new policies, methodologies and technologies. For example, at a meeting of ministerial delegations and representatives of donor organisations for the West Asian and North African countries on opportunities for sustainable investment in rain-fed areas held in 2001, the importance of developing and implementing appropriate drought policies and plans was emphasised as an urgent need.[4] Adopting a regional approach to drought management and preparedness was identified as critical to this region, allowing governments that possess experience with drought policies and preparedness to share it with others through regional and global networks.

Drought planning is an integral part of drought policy. The objectives of drought planning will, of course, vary between countries and should reflect unique physical, environmental, socioeconomic and political characteristics. A generic set of planning objectives has been developed that could be considered as part of a national, state/provincial, or regional planning effort.[5] These planning objectives have been followed or modified by numerous governments at various levels in the United States and elsewhere since the 10-step drought planning process[6] was originally developed. For example, the process has been followed in Brazil, Cyprus, Morocco and will likely be applied throughout the North Africa/Mid-East region. These objectives are set out below:

- Collect, analyse and disseminate drought-related information in a timely and systematic manner.
- Establish criteria for declaring drought and triggering various mitigation and response activities.
- Provide an organisational structure that assures information flow between and within levels of government, as well as with non-governmental organisations, and define the duties and responsibilities of all agencies with respect to drought.

- Maintain a current inventory of drought assistance and mitigation programs used in assessing and responding to drought emergencies, and provide a set of appropriate action recommendations.

- Identify drought-prone areas and vulnerable sectors, population groups and environments.

- Identify mitigation actions that can be taken to address vulnerabilities and reduce drought impacts.

- Provide a mechanism to ensure timely and accurate assessment of drought's impacts on agriculture, livestock production, industry, municipalities, wildlife, health and other areas, as well as specific population groups.

- Keep the public informed of current conditions and mitigation and response actions by providing accurate, timely information to media in print and electronic form.

- Establish and pursue a strategy to remove obstacles to the equitable allocation of water during shortages and provide incentives to encourage water conservation.

- Establish a set of procedures to continually evaluate, exercise or test, and periodically revise the plan so it will remain responsive to the needs of the people and government ministries.

Drought plans in which mitigation is a key element should have three principal components: monitoring, early warning and prediction; risk and impact assessment; and mitigation and response. A description of each of these components follows.

Drought monitoring, early warning and prediction

Effective drought early warning systems are an integral part of efforts worldwide to improve drought preparedness. Timely and reliable data and information must be the cornerstone of effective drought policies and plans. Monitoring drought presents some unique challenges because of drought's characteristics. In addition, several types of drought exist, and the factors or parameters that define drought will differ from one type to another. For example, meteorological drought is principally defined by a deficiency of precipitation from expected or 'normal' over an extended period of time, while agricultural drought is best characterised by deficiencies in soil moisture. This parameter is a critical factor in defining crop production potential. Hydrological drought, on the other hand, is best defined by deficiencies in surface and sub-surface water supplies (that is, reservoir, lake and ground water levels; stream-flow; and snow-pack), and its impacts generally lag the occurrence of meteorological and agricultural drought. These types of drought may coexist or may occur separately.

An expert group meeting on early warning systems for drought preparedness, sponsored by the World Meteorological Organisation and others, recently examined the status, shortcomings and needs of drought early warning systems, and made recommendations on how these systems can help in achieving a greater level of drought preparedness.[7] This meeting was organised as part of World Meteorological Organisation's contribution to the UNCCD meeting in Bonn, Germany, in December 2000. The proceedings of this meeting documented recent efforts in drought early warning systems in countries such as Brazil, China, Hungary, India, Nigeria, South Africa and the United States, but also noted the activities of regional drought monitoring centres in eastern and southern Africa and efforts in West Asia and North Africa. Shortcomings of current drought early warning systems were noted in the following areas:

- *data networks*—inadequate density and data quality of meteorological, and hydrological networks and lack of data networks on all major climate and water supply parameters
- *data-sharing*—inadequate data-sharing between government agencies and the high cost of data limit the application of data in drought preparedness, mitigation and response
- *early warning system products*—data and information products are often not user friendly and users are often not trained in the application of this information to decision-making
- *drought forecasts*—unreliable seasonal forecasts and the lack of specificity of information provided by forecasts limit the use of this information by farmers and others

- *drought monitoring tools*—inadequate indices for detecting the early onset and end of drought, although the Standardised Precipitation Index was cited as an important new monitoring tool to detect the early emergence of drought

- *integrated drought/climate monitoring*—drought monitoring systems should be integrated and based on multiple indicators to fully understand drought magnitude, spatial extent and impacts

- *impact assessment methodology*—lack of impact assessment methodology hinders impact estimates and the activation of mitigation and response programs

- *delivery systems*—data and information on emerging drought conditions, seasonal forecasts, and other products are often not delivered to users in a timely manner

- *global early warning system*—no historical drought data base exists and there is no global drought assessment product that is based on one or two key indicators, which could be helpful to international organisations, non-governmental organisations and others.

Participants of the expert group meeting on drought early warning systems made several recommendations. Those recommendations that pertained directly to early warning systems were that these systems should be considered an integral part of drought preparedness and mitigation plans, and that priority should be given to improving existing observation networks and establishing new meteorological, agricultural and hydrological networks.

A trend towards establishment of national and regional drought monitoring centres is apparent. For example, the regional drought monitoring centres in eastern and southern Africa have had a significant impact on the collection and dissemination of drought forecasts/outlooks and early warning information to diverse users throughout these regions since their formation a decade ago.[8] The seasonal precipitation outlooks provide users with broad regional patterns several months in advance. During periods with a strong El Niño signal (that is, higher probability of drought conditions in eastern Australia and southern Africa), the value of this information increases significantly for agriculture and other weather-sensitive sectors. Discussions regarding the establishment of other regional drought centres in other regions are ongoing. For example, a regional drought centre with a broader mission has been proposed by UNESCO following an international drought conference in South Africa in September 1999. The challenge is to link these activities closely with national drought policy and preparedness efforts in these regions.

Risk and impact assessment

Drought impacts cut across many sectors and across normal divisions of responsibility of local, state and federal agencies. These impacts have been classified by Wilhite and Vanyarkho.[9] Risk is defined by both the exposure of a location to the drought hazard and the vulnerability of that location to periods of drought-induced water shortages.[10] Information on drought impacts and their causes is crucial for reducing risk before drought occurs and for appropriate responses during drought. As part of a drought planning process, risk assessment should be undertaken by technical specialists and members of stakeholder groups that understand those economic sectors, social groups and ecosystems most at risk from drought.

An approach in accomplishing this risk assessment that has been effective in the United States is to create a series of working groups as a part of the drought planning process. These working groups will assess sectors, population groups and ecosystems most at risk, and identify appropriate and reasonable mitigation measures to address these risks. The number of working groups established varies considerably between states. This process has been widely used in the United States. This process is applied through a methodology for assessing and reducing the risks associated with drought. This methodology was completed recently through a collaboration between the NDMC and the Western Drought Coordination Council's Mitigation and Response Working Group,[11] and is available on the NDMC's web site at http://drought.unl.edu. This guide focuses on identifying and ranking drought impacts, determining their underlying causes, and choosing actions to address the underlying causes. This methodology can be employed by each of the working groups.

Mitigation and response

Mitigation is defined in several ways in the natural hazards literature. Hy and Waugh[12] referred to mitigation as activities that reduce the degree of long-term risk to human life and property. These actions normally include insurance strategies, the adoption of building codes, land-use management, risk mapping, tax incentives and disincentives, and diversification. Drought is not often directly responsible for loss of life and its impacts are largely non-structural. Therefore, this definition is not appropriate in this case. The previously stated definition for mitigation in this chapter is short- and long-term actions, programs or policies implemented during and in advance of drought that reduce the degree of risk to human life, property and productive capacity.

Mitigation needs to focus on a range of levels from micro to macro levels. Davies[13] has classified these levels as national, local government, community and household. Wilhite[14] has documented mitigation actions employed by states in the United States through a survey conducted in the early 1990s. Certainly, the range of alternatives would be greater if this survey were duplicated now since much of the country has been in severe to extreme drought conditions since 1996. The activities identified were diverse, reflecting regional differences in impacts, legal and institutional constraints, and institutional arrangements associated with drought plans. These actions represent a full range of possible mitigative actions, from monitoring and assessment programs, to the development of drought contingency plans. Some of the actions included were adopted by many states, while others may have been adopted only in a single case.

Examples of international experience with drought policy and preparedness

The United States

In 1995 the Federal Emergency Management Agency estimated average annual losses because of drought in the United States to be US$6–8 billion, more than for any other natural hazard.[15] Yet the United States has typically been ill-prepared to effectively deal with the consequences of drought. Historically, the approach to drought management has been to react to the impacts of drought by offering relief to the affected area. These emergency response programs can best be characterised as too little and too late. More importantly, drought relief does little if anything to reduce the vulnerability of the affected area to future drought events. Improving drought management will require a new paradigm, one that encourages preparedness and mitigation through the application of the principles of risk management.

There are several critical points to note about drought in the United States. First, drought occurs somewhere in the United States every year. Second, the percent area affected is highly variable from year to year, but drought years are often clustered, as in the 1930s, 1950s, late 1980s and early 1990s, and the late 1990s and early 2000s. Third, the worst year on record in terms of percent area affected was 1934, when about 65 per cent of the country was in severe to extreme drought. More recent severe drought episodes have generally been in the 40 per cent range, as was the case in 2002. Finally, no trend in the area affected is noticeable. However, impacts associated with drought in the country have increased substantially in magnitude and complexity. The implication is that vulnerability to drought is increasing.

Since 1996 widespread and severe drought conditions have occurred throughout the United States, and have raised serious concerns about continuing vulnerability to extended periods of drought-induced water shortages because of the complexity and magnitude of impacts. Many parts of the country have experienced several consecutive years of drought during this time period. Although it is not unusual for multiple drought years to occur in the drier western states, the occurrence of consecutive drought years in the east is unusual. For example, south-eastern states such as Georgia, Florida and South Carolina have experienced from three to five consecutive drought years during this time period.

Most recently, drought conditions during the period 2000–03 affected large portions of the eastern and western states. Impacts on public water supplies, agriculture, forests, transportation, energy production, recreation and tourism, and the environment (for example, fisheries, soil erosion, incidence of wild fires) have been substantial and have drawn considerable attention from elected officials and the media, providing additional fuel for the growing debate regarding the lack of a national drought policy and a coordinated response effort between federal, state, local and tribal governments.

State-level drought planning

There has been a remarkable increase in the number of states with drought plans during the past two decades. In 1982, only three states had drought plans in place. In 2003, 36 states had developed plans and five states were at various stages of plan development. The growth in the number of states with drought plans suggests an increased concern at that level about the potential impact of extended water shortages and an attempt to address those concerns through planning. The rapid adoption of drought plans by states is also a clear indication of their benefits.

Initially, drought plans largely focused on response efforts; today the trend in the United States is for states to place greater emphasis on mitigation as the fundamental element of a drought plan. Initially, states were slow to develop drought plans because the planning process was unfamiliar. With the development of drought planning models[16] and the availability of a greater number of drought plans for comparison, drought planning has become a less mysterious process for states. As states initiate the planning process, one of their first actions is to study the drought plans of other states to compare methodology and organisational structure.

Many US states have followed to a considerable degree the planning methodology outlined by Wilhite[17] and Wilhite *et al.*[18] in the development of a plan. This methodology has also been followed by tribal and local governments. Even if this methodology was not followed directly, state plans that borrow significantly from other states end up with a similar organisational structure to that proposed in the 10-step planning process.

With the tremendous advances in drought planning at the state level in recent years, it should come as no surprise that states have been extremely frustrated and dissatisfied with the lack of progress at the federal level. Early into the 1995–96 drought, the lack of leadership and coordination at the federal level quickly became obvious and continued with subsequent drought episodes. Recent initiatives towards development of a national drought policy are aimed at reducing or eliminating those frustrations.

National drought policy

Calls for action on drought policy and plan development in the United States date back to at least the late 1970s. The growing number of calls for action has resulted primarily from the inability of the federal government to adequately address the spiralling impacts associated with drought through the reactive, crisis management approach. This approach has relied on ad hoc inter-agency committees that are quickly disbanded following termination of the drought event. The lessons of these response efforts have quickly been forgotten and the failures of these efforts are subsequently repeated with the next event.

Several regional and national drought-related initiatives occurred as a result of widespread drought conditions in the United States during the period from 1996 to 1998. These initiatives lead to the passing of the National Drought Policy Act of 1998, resulting in the formation of the National Drought Policy Commission (NDPC) to 'provide advice and recommendations on creation of an integrated, co-ordinated Federal policy designed to prepare for and respond to serious drought emergencies.' The NDPC's report, submitted to Congress and the president in May 2000, recommended that the United States establish a national drought policy emphasising preparedness.[19] The goals of this policy would be to:

1   incorporate planning, implementation of plans and proactive mitigation measures, risk management, resource stewardship, environmental considerations, and public education as key elements of an effective national drought policy

2   improve collaboration among scientists and managers to enhance observation networks, monitoring, prediction, information delivery, and applied research and to foster public understanding of and preparedness for drought

3   develop and incorporate comprehensive insurance and financial strategies into drought preparedness plans

4   maintain a safety net of emergency relief that emphasises sound stewardship of natural resources and self-help

5   coordinate drought programs and resources effectively, efficiently and in a customer-oriented manner.

The legacy of the 1996 and subsequent droughts is not likely to be their impacts, but rather the policy initiatives that occurred in the post-drought period.[20] These initiatives appear to be changing the way droughts are viewed, and they may change the way droughts are managed in the United States. The real question at this point is whether these changes will result in permanent and substantive modifications in the way government entities deal with drought. The National Drought Preparedness Act of 2002 was introduced in May 2002 in both the US Senate and House of Representatives. The goal of this Bill is to develop a national drought policy that emphasises risk management through improved levels of monitoring, preparedness and mitigation. This Bill has strong support from the states and bipartisan support in Congress. The National Drought Preparedness Act was not passed by Congress in 2002. It is expected that this Bill will be considered by the US Congress in 2003. Now, more than at any time in the history of drought management in the United States, the country is at a critical crossroads for drought policy. Will it continue down the road of crisis management or move towards risk management?

Progress in Sub-Saharan Africa

In Sub-Saharan Africa, drought is a major threat to sustainable livelihoods, in particular in dryland areas of arid and semiarid regions.[21] Recent drought events have had serious economic, social and environmental consequences, and have resulted in land degradation, human migrations or relocations, famine, diseases and loss of human life.[22] In 1986, approximately 185 million people living in the dryland areas of Africa were at risk and 30 million were immediately threatened.[23] Drought has affected nearly all of the countries in western, eastern and southern Africa in the past two decades, and in many cases on more than one occasion. These droughts have resulted in a recurring deficiency of food supplies and the need for interventions by governments and international donors to alleviate food shortages to avert major losses of human life. For example, the 1991–92 drought in southern Africa resulted in a deficit of more than 6.7 million tons of cereal supplies, which affected more than 20 million people.[24] Past drought response programs have been reactive and have done little, if anything, to reduce the impacts of future droughts.

In 1997, a UNDP/UNSO project was initiated to assess the status of drought preparedness and mitigation activities in selected sub-Saharan African countries.[25] Three main questions were addressed in this assessment. First, what is the status of drought preparedness (that is, institutional capacity) within each country? Second, what constraints exist with regard to policy and plan development? Third, what are the primary drought policy and planning needs? The conclusions summarised here are drawn from 11 of the most drought-prone southern African countries: Angola, Botswana, Lesotho, Malawi, Mauritius, Mozambique, Namibia, South Africa, Swaziland, Zambia and Zimbabwe.

Common themes on the current status of drought preparedness and institutional capacity in sub-Saharan Africa included the following:

- there is no permanent government body to deal with drought issues
- drought response is often coordinated through natural disaster authorities
- drought relief is directed towards human relief, protection of key assets and recovery
- post-drought evaluation of response is not usually undertaken
- formal drought plans are rare and mainly directed at response actions
- drought and famine early warning systems commonly coexist
- vulnerability assessments often exist for sectors, groups and areas at risk
- mitigation actions focus on economic diversification and poverty reduction
- drought management is increasingly viewed as part of the development process
- drought policies are usually lacking.

Botswana and South Africa clearly stand apart from the other countries included in this assessment in terms of their experiences and current status of drought planning. Although Botswana does not have an identified drought policy and plan, it has had a long history with various types of drought programs. Drought preparedness planning is part of development planning and the institutional structure is well defined, with local involvement at the district level. In South Africa, the National Consultative Drought Forum was established in 1992 and composed of representatives of government, church organisations, trade unions and NGOs. The Forum led to a shift from an exclusive emphasis on commercial farmers to a more comprehensive program that includes rural farmers, rural poor and farm workers. Policy changes included greater equity for recipients of assistance. Drought policies have increasingly focused on improving levels of self-reliance, reducing risk in the agricultural sector and stabilising income. The National Drought Management Committee was established in 1995 with similar structures at the provincial and local levels of government. The primary objectives of this committee were to develop national disaster management policy, propose and review new legislation, promote community participation in disaster management, promote the establishment of an integrated disaster information system, and ensure risk reduction at the national level. In 2002 the South African government was looking at additional drought policy revisions.[26]

No drought policy or plan currently exists in Angola, Lesotho, Malawi, Mauritius, Mozambique, Namibia, Swaziland, Zambia or Zimbabwe, although some infrastructure does exist in most of these countries to respond to drought conditions. This has usually been only on a reactive or ad hoc crisis management basis. Two early warning systems are often in place, one focusing on monitoring climate and water supply conditions, and the other emphasising issues associated with food security. Vulnerable sectors, peoples or regions have been identified in many of these countries, but mitigation actions and programs have been limited. Response actions are generally a joint effort between government authorities, donors, NGOs and others. Most of the countries mentioned above have made considerable progress in coordinating and incorporating the capacities of donors and NGOs in drought-related emergency responses. For example, in Swaziland, a consortium of NGOs has been identified to address the needs of vulnerable population groups.

Numerous constraints to drought policy and plan development were identified in the country reports. These included:

- poor quality of meteorological networks
- minimal understanding of drought impacts
- lack of institutional capacity
- low level of involvement by NGOs in drought management
- lack of understanding of household vulnerability
- inadequate financial resources for drought management and human resources development
- need for expanded extension services
- inequitable access to land
- limited coordination between government agencies
- reduced response/mitigation capability due to lack of drought policy and plan.

Future drought policy and planning needs were also identified in the country reports. Many of these needs are aimed at addressing the constraints referred to previously. In many countries it was reported that recommendations on drought policies and specific mitigation actions had been made in government reports or as a result of workshops focused on future drought planning and response needs. In many cases, however, these recommendations have not been implemented. For example, Namibia has developed a series of drought policy recommendations based on the elements of the 10-step drought planning process developed by Wilhite.[27] The goal of the Namibian policy is to develop an efficient, equitable and sustainable approach to drought management that shifts responsibility from government to the farmer. The tenets of that policy are to (1) ensure household food security is not compromised by drought; (2) encourage and help farmers adopt a self-reliant approach to drought risk; (3) preserve reproductive capacity of the national livestock herd during drought; (4) ensure a continuous supply of potable water to communities and livestock; (5) prevent degradation of the natural resource base; (6) enable rural inhabitants and the agricultural sector to recover quickly following drought; (7) ensure the health status of all Namibians; and (8) finance drought relief programs efficiently by establishing an independent and permanent national drought fund.

Increased inter-agency coordination and the need to enhance institutional capacity were also considered important. Other needs identified included creation of a permanent national drought fund in support of mitigation and response measures, expanded meteorological networks and more comprehensive early warning systems, improved vulnerability assessments and vulnerability tracking systems, increased community participation and involvement, expanded NGO involvement in drought management, and the development of strategic grain reserves.

As expected, there is a wide range of institutional capacity to respond to drought emergencies in southern Africa. Although some countries have an organisational structure in place to coordinate the actions of government at various levels, as well as those of donors and non-governmental organisations (NGOs), most have not developed a permanent institutional capacity. One of the common problems with drought and other natural hazards is maintaining interest in planning beyond the relatively short window of opportunity that follows the event, given the on-again, off-again nature of drought. Interest in drought planning quickly wanes in the post-drought period when precipitation conditions have returned to normal or above-normal levels. The challenge is to break this cycle by developing and implementing comprehensive drought preparedness plans that emphasise risk management.

Australia

As outlined in Chapter 3, Australia officially adopted a risk management approach to drought in 1992. This policy included many of the characteristics outlined above, with its focus on increased research and development on climate patterns, an emphasis on self-reliance by agricultural producers and the intention to move away from ad hoc responses to drought. As illustrated elsewhere in this volume, the implementation of the National Drought Policy has not always met its objectives; however, it is a step in the right direction. It also highlights the difficulties governments can face in implementing a preparedness approach to drought, even in comparatively wealthy countries in which drought is a recurring phenomenon.

Global drought preparedness network

Because of increasing concern over the escalating impacts of drought and society's inability to effectively respond to these events in the past, developing and developed countries are now placing greater emphasis on the development of national policies and plans that stress the principles of risk management. Global initiatives, such as the UN Convention to Combat Desertification (UNCCD), are emphasising the importance of improving drought early warning systems and seasonal climate forecasts, and developing drought preparedness plans.

The National Drought Mitigation Center at the University of Nebraska at Lincoln is working in partnership with key UN agencies, US federal agencies, NGOs, and appropriate regional and national institutions to build a Global Drought Preparedness Network that will promote the concepts of drought preparedness and mitigation with the goal of building greater institutional capacity to cope with future episodes of drought. A Global Drought Preparedness Network could provide the opportunity for nations and regions to share experiences and lessons learned (successes and failures) through a virtual network of regional networks; for example, information on drought policies, emergency response measures, mitigation actions, planning methodologies, stakeholder involvement, early warning systems, automated meteorological networks, the use of climate indices for assessment and triggers for mitigation and response, impact assessment methodologies, demand reduction/water supply augmentation programs and technologies, and procedures for addressing environmental conflicts.

Conclusion

As this book has argued for Australia, there is a need internationally to build awareness of drought as a normal part of climate. It is often considered to be a rare and random event—thus the lack of emphasis on preparedness and mitigation. Improved understanding of the different types of drought and the need for multiple definitions and climatic/water supply indicators that are appropriate to various sectors, applications and regions is a critical part of this awareness-building process.

A second challenge is to erase misunderstandings about drought and society's capacity to mitigate its effects. Many people consider drought to be purely a physical phenomenon. We may ask, if drought is a natural event, what control do we have over its occurrence and the impacts that result? Drought originates from a deficiency of precipitation over an extended period of time. The frequency or probability of occurrence of these deficiencies varies spatially and represents a location's exposure to the occurrence of drought. Some regions have greater exposure than others, and we do not have the capacity to alter that exposure.

As with other natural hazards, drought has both a physical and a social component. It is the social factors, in combination with our exposure, that determines risk to society. Some of the social factors that determine our vulnerability are level of development, population growth and its changing distribution, demographic characteristics, demands on water and other natural resources, government policies (sustainable versus non-sustainable resource management), technological changes, social behaviour, and trends in environmental awareness and concerns. It is obvious that well-conceived policies, preparedness plans and mitigation programs can greatly reduce societal vulnerability and therefore the risks associated with drought.

A fourth challenge is to convince policy and other decision-makers that investments in mitigation are more cost-effective than post-impact assistance or relief programs. Evidence from around the world, although sketchy, illustrates that there is an escalating trend of losses associated with drought in both developing and developed countries. Also, the complexity of impacts is increasing. It seems clear that investments in preparedness and mitigation will pay large dividends in reducing the impacts of drought. A growing number of countries are realising the potential advantages of drought-planning. Governments are formulating policies and plans that address many of the deficiencies noted from previous response efforts that were largely reactive. Most of the progress made in drought preparedness and mitigation has been accomplished in the past decade or so. Although the road ahead will be difficult and the learning curve steep, the potential rewards are numerous. The crisis management approach of responding to drought has existed for many decades, and is ingrained in our cultures and reflected in our institutions. Movement from crisis to risk management will certainly require a paradigm shift. The victims of drought have become accustomed to government assistance programs. In many instances, these misguided and misdirected government programs and policies have promoted the non-sustainable use of natural resources. Many governments have now come to realise that drought response in the form of emergency assistance programs only reinforces poor or non-sustainable actions and decreases self-reliance.

Internationally, progress in drought preparedness is accelerating as knowledge of drought planning tools become more widely known and drought impacts increase in magnitude and complexity. Many regional efforts are under way to provide greater emphasis on drought policies and plans. Recent international and regional drought conferences and workshops are good examples of this growing momentum. As nations continue to build institutional capacity to cope with drought, it is imperative that these lessons learned are shared with others. Working individually, many nations and regions will be unable to improve drought coping capacity. Collectively, working through global and regional partnerships, we can achieve the goal of reducing the magnitude of economic, environmental and social impacts associated with drought in the twenty-first century.

---

1   Wilhite 2000, pp. 3–4.
2   O'Meagher *et al.* 2000, p. 115.
3   UNCCD 1999, p. 14.
4   Rabat Declaration 2001, p. 1.
5   Wilhite *et al.* 2000b, p. 697.
6   Wilhite 1991, p. 29.
7   Wilhite *et al.* 2000, p. 177.
8   Ambenje 2000, p. 131.
9   Wilhite and Vanyarkho 2000, p. 248.
10 Blaikie *et al.* 1994, p. 9.
11 Knutson *et al.* 1998, p. 1.
12 Hy and Waugh 1990, p. 19.
13 Davies 2000, p. 10.
14 Wilhite 1997, p. 961.
15 Federal Emergency Management Agency 1995, p. 2.
16 Wilhite 1991, p. 29; Wilhite *et al.* 2000b, p. 697.
17 Wilhite 1991, p. 29.

18 Wilhite *et al.* 2000b, p. 697.
19 National Drought Policy Commission 2000, p. 6.
20 Wilhite 2001, p. 20.
21 Glantz 1987, p. 43.
22 UNDP/UNSO 2000, p. 3.
23 Dinar and Keck 2000, p. 137.
24 Dinar and Keck 2000, p. 138.
25 UNDP/UNSO 2000, p. 3.
26 Monnik 2000, p. 48.
27 Wilhite, 1991, p. 29; Wilhite *et al.* 2000b, p. 697.

10

Beyond drought in Australia:
The way forward

Linda Courtenay Botterill

This collection of essays has provided a multi-disciplinary perspective on drought in Australia. It has brought together the diverse views of experts from a range of fields. Perhaps surprisingly, consistent themes have emerged from the work of our contributors.

The overriding theme that has emerged from the research has been the need for Australians and specifically farmers and the rural policy community to be comfortable with 'being Australian'. In this context we mean developing a deep understanding and empathy for the realities of our biophysical environment and climate systems. Once this is achieved, we will be able to move away from an idealised, essentially European image of climate, weather and agriculture. Part of this understanding will come from improved climate literacy. This means living with a highly variable climate—and recognising that the concept of 'average' rainfall is essentially a statistical construct that bears little resemblance to most seasons. This approach suggests that the notion of 'drought' may be meaningless in an environment in which extremes are the norm, particularly as the term is so value laden and evocative of unexpected disaster.

As our authors have shown, drought is not easily defined. It can be meteorological, hydrological, agricultural or social. In the

Australian context, we would add the need to consider the environmental impact of human responses to low rainfall. Each of these facets of drought can suggest different policy responses. Each policy response has its own objectives, and policy-makers are faced with the challenge of balancing these different, sometimes competing goals. The challenge in a liberal democracy is to frame policy that deals with the complexity of the issues surrounding drought and strikes an appropriate balance between the interests of competing stakeholders. In essence, it is drought's impact on human activity that is its defining characteristic and this means that at its heart drought itself is a political event. When drought hits the public agenda, politicians, the media, lobby groups and policy-makers are thrown into a high pressure, highly emotive environment in which the policy process must proceed under unusual levels of scrutiny and political pressure. The inevitable trade-offs involved in any policy solution can be highlighted and subject to considerable criticism because the stakes are so high and the immediate human costs are very real.

I will return to the three main themes of this book later in this chapter. First, I would like to highlight some of the other interesting points that our authors have raised. Daniela Stehlik and Mark Stafford Smith have both raised the issue of resilience and the need to develop policies that enhance the resilience of our human communities and our ecosystems to strengthen their capacity to withstand drought. This is consistent with the National Drought Policy approach of self-reliance, but emphasises that systems, both human and natural, need to be understood to ensure that resilience is truly supported by policy measures. A good starting point is for policy-makers to question some of their assumptions about the way systems operate and to learn from the feedback from social and biophysical systems. This means, for example, not assuming that adversity brings about community cohesion and not only focusing on 'fast' variables when assessing the impact of drought.

Peter Hayman and Peter Cox raise the idea that drought can be a metaphor for rural suffering and point to our tendency to personify drought. Åsa Wahlquist makes a similar observation when she reports our tendency to talk about drought in the language of war and conflict. This habit of describing drought as 'the enemy' is not a uniquely Australian phenomenon,[1] but it does undermine our capacity to regard drought as a normal part of our environment. As West and Smith have argued, 'Notwithstanding the lessons that nature has taught us, droughts are consistently defined as unexpectedly severe in their intensity or duration'.[2]

The agrarian overtones that appear in public discourse as drought worsens, and the rise of appeals such as Farmhand, suggest that the broader community is sympathetic to the plight of farmers affected by drought. Interestingly, Stehlik's research suggests that farmers feel abandoned and misunderstood by the broader community. The mismatch between the perception of the media and politicians about the public's response to the drought, and that of those experiencing the impact of drought first-hand, is as disturbing as it is poignant. Urban Australia arranges appeals, metropolitan newspapers run sympathetic editorials and, as Wahlquist argues in this volume, their stories are biased to those of hardship—and yet Stehlik writes (Chapter 5) that 'for many of our respondents, the lack of understanding about their experience and the hardship they were undergoing, was evidenced by what they saw as a "rejection" by others of their "whingeing" and "complaining"'. It will be interesting to see how the prospect of increased interest in drought policy by urban voters, raised by Hayman and Cox and by Neil Inall in Wahlquist's chapter, is received in rural Australia.

Learning to be Australian

In *The Future Eaters*, Tim Flannery makes the case that we need to learn to live on this continent in a way that is sympathetic to it strengths and limitations. To a large extent, the history of agriculture in Australia has been characterised by our efforts to turn this dry country with its poor, ancient soils into another Europe. With European style-agriculture have come expectations of climate that has predictable seasons and fits a pattern of 'normal' behaviour around which we can plan our agricultural activities. However, the European climate is essentially predictable and based on an annual cycle. By contrast, Flannery notes that 'Australia is the only continent on Earth where the overwhelming influence on climate is a non-annual climatic change'.[3]

As this book was completed in 2003, we were experiencing a widespread drought that reduced our national income, led to water restrictions in many of our major cities and caused considerable hardship for farm families across Australia. In many places, the Commonwealth government declared the drought to be an 'exceptional circumstance' and provided relief to the farm family, farm businesses and non-farm small businesses. However, the terminology of exceptional circumstances still hints that extreme dry spells are aberrations to which governments need to respond. The frequency with which the exceptional circumstances provisions have been triggered since their introduction in 1993 suggests that we have not moved far from the declaration of a natural disaster when drought occurs. As Åsa Wahlquist describes in her chapter, this view of drought as an enemy, not an unpredictable neighbour, is perpetuated in the media and the public mind.

Part of 'learning to be Australian' perhaps also includes recognising the historical and cultural settings within which agriculture operates. While we have certain biophysical and climate conditions with which to work, we also have social and cultural settings that need to be considered. Attachment to the family farm as the best form of agriculture for Australia is widespread, as is an attachment to rural communities and the rural way of life as somehow being quintessentially Australian. Celebrations such as the bicentenary, the Sydney Olympics, the Centenary of Federation and the Year of the Outback reflect the resonance of rural imagery across the broader Australian community. As Heathcote has argued 'In any catastrophe, public sympathy goes out to the victims, but when those victims are the sons of the soil, on the margins of the good earth, struggling to give us our daily bread, the emotional response is tremendous and objectivity is often left behind'.[4] This emotional response sets the tone for the policy debate during drought events. It also suggests that policy approaches which may lead to a future of large-scale corporate farming are not politically palatable.

Policy-makers therefore face the challenge of ensuring that drought policy remains compatible with the government's structural adjustment objectives for the farm sector, while meeting the welfare needs of drought-affected farm families. The balance between these often-competing aims has changed over the years with recent policies focusing on farming as an economic undertaking and an emphasis on structural adjustment rather than income support.[5] The idea that farming is a business is not new. In a textbook claimed by its author to be the first on Australian agriculture, W. Catton Grasby wrote in 1912 that 'farming is very largely a matter of business'.[6] Since 1992, Commonwealth Ministers responsible for agriculture have pressed the point, with references to 'farm business managers'[7] and the 'transition from family farm to family farm business'.[8] This delineation has been reflected in the removal of farm welfare provisions from programs, the objective of which is facilitating the process of structural adjustment in agriculture.

Drought can exacerbate and bring to the fore otherwise hidden rural hardship. It can also reveal bad farm management. It must be recognised that for every farmer featured in the media in difficulty, there are others who are managing the impact of drought without requiring assistance. On equity grounds alone, it is important that drought relief not perpetuate otherwise economically and environmentally unsustainable businesses or reward poor farm managers at the expense of the better prepared. Although a focus on risk management did not become a part of drought policy until 1992, concerns along these lines were raised as early as the mid-1960s. In 1965, a Select Committee of the NSW Legislative Assembly reporting on drought relief included the following in its recommendations to the State government:

> *Your committee feels in its recommendation, that the measures of assistance which we believe would assist are in the form of incentives, not hand-outs. Hand-outs should be discouraged because this sometimes comes to the point where the least deserving receives the most assistance, and the man who has been a good manager and done a great deal to prepare for some such catastrophe as drought can be the one who will not receive any form of assistance.*[9]

This is not a trivial issue. The amounts of money spent on drought relief have been substantial in recent years, and it is therefore not unreasonable to consider the appropriate division of responsibility between farmers and governments for drought management. The National Drought Policy explicitly set out roles for government and farmers in July 1992, indicating that 'farmers will have to assume greater responsibility for managing risks arising from climatic variability', while 'Government will create the overall environment which is conducive to this whole farm planning and risk management approach' and 'During severe downturns … act to preserve the social and physical resource base of rural Australia'.[10] Achieving policy settings that are conducive to this sharing of responsibility is challenging.

Complexity and policy trade-offs

This volume has brought together a range of disciplinary perspectives on drought and its impact. We have set out to examine the issue from the macro level of considering drought in the context of the landscape and its history, through to the very personal experiences of those for whom drought is a lived event. In doing this, we seek to demonstrate that none of these perspectives is irrelevant to the policy debate and that balancing competing interests is not an easy task. Two features are worth highlighting—one is the ongoing impact of agrarianism among farming communities and the other is managing the unintended consequences of policy interventions.

In recent years, the focus of policy-makers has been on the economic performance of the farm sector and policies have been based on the provision of appropriate economic incentives to influence farmer behaviour. It is easy to dismiss as unimportant non-economic drivers of behaviour and to suggest that there is no place in rational policy debate for consideration of apparently irrational motivations. As Stehlik's chapter in this volume illustrates so vividly, drought affects the lives of real people. Real people often appear to act quite differently from the rational utility-maximising agents of economic theory. They measure their utility in their own way and against their own values. The implication of this observation for policy is that, in order to understand and perhaps influence farmer behaviour, it may be useful to look beyond the economic model and consider other perspectives. It was suggested above that the bush is an important part of the Australian self-image. The images that the 'rural' evokes are associated with agrarian ideals, or in the Australian context, countrymindedness. This emotional attachment to farming cannot be ignored by the policy process—it is very evident in media discourse and political rhetoric during drought, and this in turn feeds into the political process. Disregarding its potency can also impact on the efficacy of government programs, leading to policy failure.[11]

In 1992, Commonwealth and State Ministers agreed that the National Drought Policy would be 'based on principles of sustainable development, risk management, productivity growth and structural adjustment in the farm sector'.[12] Since then, other attempts have been made at developing an 'Integrated Rural Policy Package'[13] and a 'Business plan for Australian agriculture',[14] suggesting a desire on the part of policy-makers to eliminate inconsistencies and develop a cohesive policy approach. The National Drought Policy called for the elimination of transaction-based subsidies such as fodder and transport subsidies, as they were considered to encourage delayed de-stocking by farmers, resulting in environmental damage. Putting aside the fact that NSW and Queensland continued to make these subsidies available during the 2002–03 drought, this signalled an attempt to consider the unintended consequences of drought relief measures.

Shades of drought—policy responses in the face of uncertainty

Not all droughts are created equal, and it is not clear at the beginning of a dry spell whether the current circumstance will be manageable or require government intervention. Our understanding of the El Niño Southern Oscillation Cycle is helping us in assessing the likelihood of a prolonged drought, but our ability to predict severe drought remains limited. It is also apparent that the impact of climate change, whether anthropogenic or naturally occurring, needs to be factored into the equation.

In recognition that droughts vary in severity, the National Drought Policy introduced in 1992 made a distinction between normal drought and 'exceptional circumstances' drought, with the differentiating element relating to whether a good farm operator could be expected to manage the impact of the event. However, the very existence of the exceptional circumstances provisions introduces a factor into the farmer's planning process. The issue of when a drought becomes exceptional is critical when, arguably significant, government support is tied to the concept. The situation is further complicated by the existence of separate state drought declaration processes that vary in their objectivity between states, and in fact some do not declare droughts at all. The state declarations allow for media headlines like 'Drought Hits 98% of NSW',[15] but do not provide an accurate picture of the severity or extent of the dry spell being described.

The National Drought Policy did not get off to a good start in this regard. 'Exceptional circumstances' were not defined in the relevant legislation or the Second Reading Speech, and the first exceptional circumstances declarations were largely subjective. Making the distinction between shades of drought has been an ongoing issue in the administration of the policy. It is not, however, a new problem. In 1967, it was argued that:

> *Perhaps we can fairly say that in many regions we can now successfully handle the 'normal' seasonal droughts, whether regularly in winter or in summer, and no matter how intense. But the great droughts, the 'old man droughts', when seasonal rains fail—and fail yet again—are cataclysmic in their effects.*[16]

In 1973, writers were discussing the impact of 'rare, unusual droughts which are not or cannot be protected against by the normal routine of farm operations'.[17]

Responding appropriately to drought is only one of a government's overall objectives for the rural sector, and the linkages between drought and other policy areas are important. Two policy objectives that have gained prominence since 1990 are the desire to support the rural sector as it adjusts to changing economic conditions and the need to preserve the natural resource base. These objectives have implications for the implementation of drought policies, which may encourage otherwise non-viable farmers to remain in farming or encourage farming practices that are environmentally unsustainable.

Perhaps the first question to be asked when constructing drought policy is whether support should be given at all, and if so, whether it is appropriate to provide it to farm businesses, farm families or both. As soon as there is government intervention, there is potential for unintended consequences elsewhere in the system, as illustrated by Stafford Smith in Chapter 7, and the costs and benefits of providing support need to be assessed. A close examination of drought policy can lead to a wider questioning of other policy approaches. For example, would drought policy be so complex and have such potential for politicisation if farm families had easier access to the general social security safety net? Farm welfare has tended to be treated as an off-shoot of structural adjustment policy in Australia. Policies to address inadequate farm incomes have been developed in the agriculture department, not the welfare portfolio, and as such, structural adjustment objectives have been incorporated in the framework of what are essentially income support programs.[18]

A further important consideration is the need to develop transitional arrangements between old and new programs. This inevitably constrains policy-makers in their choice of policy instruments and can contribute to incrementalism in the policy process. Although there is much debate about whether incrementalism delivers optimum policy outcomes,[19] it is generally recognised as the way policy is made.[20] We accept that there are sound reasons why policy cannot be developed more 'rationally'; however, we suggest policy-makers regularly question the assumptions underlying their policy approaches and the objectives inherent in the policy instruments they are employing. While incrementalism is a feature of the development of the policy means, decision-makers must not lose sight of the policy ends towards which incremental change is leading.

The impact of politics on the drought response is an ongoing issue. As I have argued elsewhere,[21] changes to drought policy during severe droughts have tended to soften the arguably more economically rational approach taken to policy at other times. The Drought Relief Payment, access to support on the basis of *prima facie* evidence and the exemption of Farmhand donations from the income test for government income support payments were all announced during exceptional circumstances droughts. The application process for exceptional circumstances support lends itself to politicisation, as farmers 'make a case' that they qualify, and the involvement of State governments, often of different party affiliations from the Commonwealth, further increases political pressures.

Elected politicians cannot disregard the impassioned calls for assistance from drought-affected farmers and their supporters. Even at the beginning of an electoral cycle, politicians at one level of government can come under pressure from those at another faced with imminent elections. Tensions are further heightened when State and Commonwealth governments are of different political persuasions and they capitalise on policy uncertainty for political purposes.

As Wahlquist describes in this volume, the media has an important role during drought in influencing the public mood and reporting the impact of the drought objectively. During the 2002–03 drought we saw the interesting phenomenon of a news organisation actively sponsoring the Farmhand appeal only four months after arguing that farmers should take advantage of the good times to 'provide a buffer for the inevitable downturn in the cycle'. The piece suggested that 'taxpayers have coughed up election-related sweeteners like the $1.9 billion being doled out to diary farmers ... and the $810 million Agriculture Advancing Australia package that was stitched together to prevent rural voters deserting the coalition'.[22] Within a relatively short space of time, the same newspaper was running an editorial with the title 'Lending hand to farmers benefits us all'.[23]

It is not unreasonable to expect governments to be responsive in a liberal democracy like Australia. It is, however, not always conducive to the development of a cohesive, sustainable policy approach. The introduction of some form of drought relief based on farmers' individual circumstances, rather than on declaration of an exceptional circumstance, could greatly diffuse the political situation and allow for the development of more considered policy responses.

Conclusion

The authors who have contributed to this book have set out to illustrate three important themes that we believe can contribute to the development of drought policy in Australia. First, we recognise the reality of the Australian climate. It is highly variable and as Australia enters the second century of its existence as a Federation, we suggest it is time that we reject constructions of climate that are inconsistent with our experience of this continent. Concepts of 'normal' and 'average' seasons that are interrupted by the aberration of drought need to be rejected in favour of an approach that accepts the inherent uncertainty or our weather patterns and seeks to work with them. Second, we have sought to highlight the complexity of drought: its economic, environmental and social impacts. This idea is not new, as previous discussions of the definition of drought have highlighted, but it is important enough to restate. Finally, we do not seek to recommend a single 'correct' course of action. As indicated, we argue that such a policy does not exist. We seek to draw attention to the factors that we see as important to the development of effective drought policy in Australia. We hope that policy-makers and stakeholders in the policy process will find this valuable in their consideration of government responses to drought.

1 Tadesse 2000, p. 141.

2 West and Smith 1996, p. 97.

3 Flannery 1994b, p. 81.

4 Heathcote 1973, p. 36.

5 Botterill 2003a.

6 Grasby 1912, p. 278.

7 Crean 1992.

8 Anderson 1997.

9 Parliament of NSW 1966, p. ix.

10 ACANZ 1992, p. 13.

11 Botterill 2001b.

12 ACANZ 1992, p. 13.

13 ARMCANZ 1997.

14 ARMCANZ 1996.

15 O'Meagher *et al.* 2000.

16 Donald 1967, p. 84.

17 Hill 1973, p. 198.

18 Botterill 2001a.

19 Lindblom 1959; Dror 1964; Lowi 1979; Burch and Wood 1983.

20 For example, see Wildavsky 1979; Hogwood and Gunn 1984; Ham and Hill 1993; Hayes 1992.

21 Botterill 2003b.

22 *The Australian* 2002a.

23 *The Australian* 2002b.

References and Bibliography

Agricultural Council of Australia and New Zealand (ACANZ) (1992) *Record and Resolutions: 138th Meeting, Mackay 24 July 1992*, Commonwealth of Australia

Agriculture and Resource Management Council of Australia and New Zealand (ARMCANZ) (1996) *Record and Resolutions: Eighth Meeting, Cairns 27 September 1996*, Commonwealth of Australia

Agriculture and Resource Management Council of Australia and New Zealand (ARMCANZ) (1997) *Record and Resolutions: Eleventh Meeting, Darwin 2 August 1997*, Canberra, Commonwealth of Australia

Agriculture and Resource Management Council of Australia and New Zealand (ARMCANZ) (1999a) *Record and Resolutions: Fifteenth Meeting, Adelaide 5 March 1999*, Canberra, Commonwealth of Australia

Agriculture and Resource Management Council of Australia and New Zealand (ARMCANZ) (1999b) *Record and Resolutions: Sixteenth Meeting, Sydney 6 August 1999*, Canberra, Commonwealth of Australia

Agriculture and Resource Management Council of Australia and New Zealand (ARMCANZ) (2001) *Record and Resolutions: Twenty-first Meeting, Darwin 17 August 2001*, Canberra, Commonwealth of Australia

Aitkin, Don (1985) '"Countrymindedness"—the spread of an idea', *Australian Cultural History* (4): 34–41

Allan, R. and R. L. Heathcote (1987) 'The 1982–83 drought in Australia' in Glantz, M. H., R. Katz and M. Krenz (Ed.), *The Societal Impacts Associated with the 1982–93 Worldwide Climate Anomalies* Boulder, Colorado, National Center for Atmospheric Research, pp. 18–23

Allan, R. J., J. A. Lindesay and D. Parker (1996) *The El Niño Southern Oscillation and Climatic Variability*, Melbourne, CSIRO Publishing

Allan, R. J. and J. A. Lindesay (1998) 'Past climates of Australasia' in Hobbs, J. E., J. A. Lindesay and H. A. Bridgman (Eds), *Climates of the Southern Continents: Present, Past and Future*, Chichester, John Wiley & Sons, pp. 207–247

Ambenje, P. G. (2000) 'Regional drought monitoring centers – The case of Eastern and Southern Africa' in Wilhite, D. A., M. V. K. Sivakumar and D. A. Wood (Eds), *Early Warning Systems for Drought Preparedness and Drought Management, Proceedings of an Expert Group Meeting Lisbon, Portugal, 5–7 September*, Geneva, Switzerland, World Meteorological Organization, pp. 131–136

Amery, the Hon Richard MP (2002) 'Howard Government Paying Little and Wanting to Pay Less', Media Release by the NSW Minister for Agriculture, 11 October 2002

Anderson, J. R. (1994) 'Risk management in Australian agriculture: An overview', in Powell, R. (Ed.), *Risk Management in Australian Agriculture*, UNE, pp. 1–12

Anderson, J. R. and J. L. Dillon (1992) 'Risk analysis in dryland farming systems', Rome, FAO

Anderson, the Hon John, MP (1997) 'Federal Government gives farm sector "AAA" rating', Media Release by Minister for Primary Industries and Energy, 14 September 1997

Arthur, J (1999) 'Dictionaries of the default country', Lingua Franca, ABC Radio National www.abc.net.au/rn/arts/ling/stories/s29287.htm

AUSLIG (1992) *The Ausmap Atlas of Australia* Cambridge, Cambridge University Press

Australian Broadcasting Corporation (2002) 'Drought Money' 9 May 2002 www.abc.net.au/rural/qld/stories/s551689.htm

Australian Bureau of Agricultural and Resource Economics (ABARE) (1991) *Australian broadacre agriculture 1990–91 and 1991–92*, Canberra, Commonwealth of Australia

Australian Bureau of Agricultural and Resource Economics (ABARE) (2001) *Alternative policy approaches to natural resource management*, Background Report to the Natural Resource Management Task Force, February 2001 Canberra, ABARE

Australian Bureau of Agricultural and Resource Economics (ABARE) (2002) *Drought continues to devastate crops*, Canberra

Barber, J. G. (In press) 'Australia's contribution to social work research and evaluation 1998–2000', *Social Work Research and Evaluation: An International Journal*.

Bardsley, P. and M. Harris (1987) 'An approach to the econometric estimation of attitudes to risk in agriculture', *Australian Journal of Agricultural Economics* **31**: 112–126

Barker, W. R. and P. J. M. Greenslade (Eds) (1982) *Evolution of the Flora and Fauna of Arid Australia*, Adelaide, Peacock Publications

Beadle, N. C. W. (1981) *Vegetation of Australia*, Cambridge, Cambridge University Press

Beck, U. (1992) *Risk society: towards a new modernity* Theory, culture & society series. London, Sage

Bensen, J. (2002) 'Letter to the editor' *Sydney Morning Herald*, 12 July 2002

Benson, C. and E. Clay (2000) 'The economic dimensions of drought in Sub-Sahara Africa' in Wilhite, D. A. (Ed.), *Drought. A Global Assessment* London, Routledge Publishers Volume 1, pp. XX–XX

Bernstein, P. L. (1996) *Against the Gods: The remarkable story of risk*, New York, John Wiley

Bigge, J. T. [1823] (1966) *Report on Agriculture and Trade in NSW* Australiana Facsimile Editions No 70 Adelaide, Libraries Board of South Australia

Blaikie, P., T. Cannon, I. Davis and B. Wisner (1994) *At Risk: Natural Hazards, People's Vulnerability, and Disasters*, London, Routledge

Blench, R. and Z. Marriage (1999) 'Drought and livestock in semi-arid Africa and southwest Asia', Overseas Development Institute Working Paper 117 London

Boadway, R. and D. E. Wildasin (1984) *Public Sector Economics* Boston, Little Brown & Co

Bond, G. E. and B. S. Wonder (1980) 'Risk attitudes amongst Australian farmers' *Australian Journal of Agricultural Economics* **24**: 16–35

Botterill, L. C. (2001a) 'Muddling through or just a muddle: Australian government responses to farm poverty 1989–1998' Unpublished PhD thesis Australian National University

Botterill, L. C. (2001b) 'Rural policy assumptions and policy failure: the case of the re-establishment grant' *Australian Journal of Public Administration* **60**(4): 13–20

Botterill, L. C. (2003a) 'Rural policy in Australia: The farm family and the farm business' in Holland, I., and J. Fleming (Eds), *Government Reformed. Values and New Political Institutions* Aldershot, Ashgate Press, pp. 89–105

Botterill, L. C. (2003b) 'Uncertain climate: The recent history of drought policy in Australia' *Australian Journal of Politics and History* **49**(1): 61–74

Botterill, L. C. and B. Chapman (2002) 'Developing equitable and affordable government responses to drought in Australia' *Australasian Political Studies Association Conference*, Australian National University, Canberra, 2–4 October 2002

Braybrooke, D. and C. E. Lindblom (1963) *A strategy of decision: policy evaluation as a social process*, New York, Free Press of Glencoe

Brownhill, Senator D. (1994) *Drought Relief Payment Bill 1994: Second Reading Debate* Senate Hansard, 13 October 1994

Bruwer, J. J. (1993) 'Drought policy in the Republic of South Africa' in Wilhite, D. A. (Ed.), *Drought Assessment, Management and Planning: Theory and Case Studies* Boston, Kluwer Academic Publishers, pp. 199–212

Bryceson, B. and D. H. White (Eds) (1992) *Proceedings of a workshop on drought and decision support, 11–13 March 1992* Canberra, Bureau of Rural Sciences

Bulis, H., D. Stehlik, G. Lawrence and I. Gray (1996) 'The shifting sands of disadvantage – A social policy analysis of drought' in Heathcote, R. L., C. C. Cuttler and J. Koetz (Eds) *NDR96 Conference on Natural Disaster Reduction Proceedings*, Surfers Paradise, Queensland, 29 September–2 October 1996

Burch, M. and B. Wood (1983) *Public policy in Britain*, Oxford, M. Robertson

Bureau of Meteorology, Australia (2003) *Climate of the 20th Century* http://www.bom.gov.au/lam/climate/levelthree/c20thc

Buxton, R., G. Brennan, J. Engleke, E. Jack and M. Stafford Smith (1995) *DroughtPlan Regional Report: Kimberley* Alice Springs, CSIRO

Clark, A. J. and Brinkley T. (2001) *Risk management for climate agriculture and policy*, Canberra, Bureau of Rural Sciences, Commonwealth of Australia

Clark, C. M. H. (Ed.) (1950) *Select Documents in Australian History 1788–1850*, Sydney, Angus and Robertson

Coakes, S. and M. Fisher (2001) 'Risk perception, farmers and biotechnology', Bureau of Rural Sciences Report prepared for AFFA Canberra

Cock, G. (2001) *Report on Impact Stage 2 1996–2000 National Property Management Planning Campaign*, Natural Resources Management Business Unit, AFFA Canberra

Cohen, K. (2002) 'Letter to editor' *Sydney Morning Herald*, 12 July 2002

Cohen, M. D., J. G. March and J. P. Olsen (1972) 'A garbage can model of organizational choice' *Administrative Science Quarterly* **17**(1): 1–265

Colls, K. and R. Whitaker (1990) *The Australian Weather Book*, Brookvale, National Book Distributors

Conacher, A. and J. Conacher (1995) *Rural Land Degradation in Australia*, Melbourne, Oxford University Press

Cox, P. G. (1996) 'Some issues in the design of agricultural decision support systems', *Agricultural Systems* **52**(2/3): 355–381

Cox, P. G., A. D. Shulman, P. E. Ridge, M. A. Foale and A. L. Garside (1995) 'An interrogative approach to system diagnosis: An invitation to the dance', *Journal for Farming Systems Research-Extension* **5**: 67–83

Crean, the Hon Simon, MP (1992) *Rural Adjustment Bill 1992: Second Reading Speech* House of Representatives Hansard, 3 November 1992

Cross, J. and M. Stafford Smith (2001) *RISKHerd: taxation policy instruments and grazing management in the rangelands*, RISKHerd Project Report No. 8 Alice Springs, CSIRO

CSIRO (2001) *Climate change projections for Australia* CSIRO Atmospheric Research, http://www.dar.csiro.au/publications/projections2001.pdf

CSIRO Division of Soils (Ed.) (1983) *Soils: an Australian viewpoint*, Melbourne, CSIRO and Academic Press

Daly, D. (1994) *Wet as a shag. Dry as a bone. Drought in a variable climate* QDPI Information series QI93028 Brisbane, Queensland Department of Primary Industries

Damasio, A. R. (1994) *Descartes' error: Emotion, reason and the human brain*, New York, Avon

Davies, S. (2000) 'Effective drought mitigation: linking micro and macro levels' in Wilhite, D. A. (Ed.), *Drought: A Global Assessment* London, Routledge Publishers Volume 2, pp 3–16

Delbridge, A., J. R. L. Bernard, D. Blair, P. Peters and S. Butler (1991) *The Macquarie Dictionary*, (Second Edition), The Macquarie Library Pty Ltd.

Department of Primary Industries and Energy (DPIE) (1996) *Mid Term Review of the Rural Adjustment Scheme: Submission by the Department of Primary Industries and Energy* Canberra

Dinar, A. and A. Keck (2000) 'Water supply variability and drought impact and mitigation in Sub-Saharan Africa. ' in Wilhite, D. A. (Ed.), *Drought: A Global Assessment* London, Routledge Publishers Volume 2, pp. 129–148

Dodson, J. R. and M. Westoby (1985) 'Are Australian ecosystems different?', *Proceedings of the Ecological Society of Australia* **14**

Donald, C. M. (1967) 'Innovation in agriculture', in Williams, D. B. (Ed.), *Agriculture in the Australian Economy* Sydney, Sydney University Press, pp. 57–86

Douglas, M. and A. Wildavsky (1982) *Risk and Culture: An Essay on the Selection of Technical and Environmental Dangers*, Berkeley, University of California Press

Drought Policy Review Task Force (DPRTF) (1990) *National Drought Policy*, Canberra, AGPS

Dracup, J., K. S. Lee and E. G. Paulson (1980) 'On the definition of droughts', *Water Resources Research* **16**(2): 297–302

Dror, Y. (1964) 'Government decision making muddling through—"Science" or "Inertia"', *Public Administration Review* **XXIV**(3): 153–157

Ellis, F. (2000) *Rural livelihoods and diversity in developing countries*, Oxford, Oxford University Press

Etzioni, A. (1967) 'Mixed-Scanning: A "Third" Approach to Decision-Making', *Public Administration Review* **XXVII**(5): 385–392

Fargher, J., B. Howard, D. Burnside and M. Andrew (2002) 'The economy of Australian rangelands', *Proceedings of Australian Rangelands Society 12th Biennial Conference*, Kalgoorlie, Australian Rangelands Society

Federal Emergency Management Agency (1995) *National Mitigation Strategy*, Washington, DC, 1995

Finucane, M. (2000) *Improving Quarantine Risk Communication: Understanding Public Risk Perceptions*, Decision Research Report 00–7 Eugene, Oregon

Flannery, T. F. (1994) *The Future Eaters: An Ecological History of the Australasian Lands and People*, Sydney, Reed Books

Foley, J. C. (1957) *Droughts in Australia. Review of records from earliest settlement to 1955* Bulletin No 43 Melbourne, Bureau of Meteorology

Forth, G. (2000) 'What is the future for Australia's declining country towns?', *Online Opinion*, 2003, http://www.onlineopinion.com.au/Aug00/Forth.htm

Freebairn, J. (2002) *Drought Connections*, Newsletter of Agribusiness Association of Australia and the Australian Agricultural and Resource Economics Society, Spring (December): 3

Freebairn, J. W. (1983) 'Drought and assistance policy', *Australian Journal of Agricultural Economics* **27**(3): 185–199

Friedel, M. H., B. D. Foran and D. M. Stafford Smith (1990) 'Where the creeks run dry or ten feet high: pastoral management in arid Australia', *Proceedings of the Ecological Society of Australia* **16**: 185–194

Fuller, S. (2002) 'Social capital is capital by more polite means: How information technology exacerbates the traditional problems of capitalism', *XV World Congress of Sociology. The Social World in the 21st Century: Ambivalent Legacies and Rising Challenges*, Brisbane, 7–13 July 2002.

Gerritsen, R. (1987) 'Why the "uncertainty"? Labor's failure to manage the "rural crisis"', *Politics* **22**(1): 47–59

Gibbs, W. J. and J. V. Maher (1957) *Droughts in Australia: Bulletin No 43* Melbourne, Commonwealth Bureau of Meteorology

Gigerenzer, G. and P. M. Todd (1999) *Simple Heuristics That Make Us Smart*, New York, Oxford University Press

Glantz, M., R. Katz and M. Krenz, (Eds) (1987) *The Societal Impacts Associated with the 1982–83 Worldwide Climate Anomalies* Boulder, Colorado, National Center for Atmospheric Research

Glantz, M. H. (1987) *Drought and Hunger in Africa: Denying Famine a Future*, Cambridge, Cambridge University Press

Glantz, M. H. (2000) 'Drought follows the plough: A cautionary note', in Wilhite, D. A. (Ed.), *Drought: A Global Assessment* London, Routledge Publishers Volume 2, pp. 285–291

Glantz, M. H., R. W. Katz and N. Nicholls (Eds) (1991) *Teleconnections Linking Worldwide Climate Anomalies*, Cambridge, Cambridge University Press

Grasby, W. C. (1912) *Principles of Australian Agriculture*, London, Macmillan & Co

Gray, I. and G. Lawrence (1996) 'Predictors of stress among Australian farmers', *Australian Journal of Social Issues* **31**(2): 173–189

Gray, I., D. Stehlik, G. Lawrence and H. Bulis (1998) 'Community, communion, and drought in rural Australia', *Journal of the International Community Development Society* **29**(1): 23–37

Greig, A., F. Lewins and K. White (Eds) (2002) *Inequality in Australia*, Cambridge, Cambridge University Press

Grimson, M. (Ed.) (1999) *Taking Control: Farmers talk about Property Management Planning*, Kingston, ACT, National Farmers' Federation

Grogan, F. O. (1968) 'The Australian Agricultural Council: A successful experiment in Commonwealth–State relations' in Hughes, C. A. (Ed.), *Readings in Australian Government*, St Lucia, University of Queensland Press: 297–317

Groves, R. H. (1994) *Australian Vegetation*, Cambridge, Cambridge University Press

Haberkorn, G., G. Hugo, M. Fisher and R. Aylward (1999) *Country Matters: Social Atlas of Rural and Regional Australia*, Canberra, Bureau of Rural Sciences

Ham, C. and M. J. Hill (1993) *The policy process in the modern capitalist state* (2nd Edition), New York, Harvester Wheatsheaf

Hammond, K. R. (1996) *Human Judgement and Social Policy. Irreducible uncertainty, inevitable error, unavoidable injustice*, New York, Oxford University Press

Hardaker, J. B., R. B. M. Huirne and J. R. Anderson (1997) *Coping with Risk in Agriculture*, Wallingford, CAB International

Hattis, T. R. (1994) 'Communicating environmental risk', *Risk Analysis* **5**: 704–760

Hayes, M. T. (1992) *Incrementalism and public policy*, New York, Longman

Hayman, P. T. (2001) 'Farmers and agricultural scientists in a variable climate', Unpublished PhD thesis, University of Western Sydney

Heathcote, R. L. (1973) 'Drought perception' in Lovett, J. V. (Ed.), *The Environmental, Economic and Social Significance of Drought*, Sydney, Angus & Robertson, pp. 17–40

Heathcote, R. L. (1994) 'Australia' in Glantz, M. H. (Ed.), *Drought Follows the Plough*, Cambridge, Cambridge University Press, pp. 91–102

Heathcote, R. L. (1998) 'Drought in Australia: Still a problem of perception?' *GeoJournal* **16**(4): 387–397

Hennessy, K. J., R. Suppiah and C. M. Page (1999) 'Australian rainfall changes, 1910–1995', *Australian Meteorological Magazine* **48**: 1–14

Hill, M. K. (1973) 'Farm management for drought mitigation', in Lovett, J. (Ed.), *The Environmental, Economic and Social Significance of Drought*, Sydney, Angus & Robertson, pp. 195–219

Hobbs, J. E. (1998) 'Present climates of Australia and New Zealand', in Hobbs, J. E., J. A. Lindesay and H. A.Bridgman (Eds), *Climates of the Southern Continents: Present, Past and Future*, Chichester, John Wiley & Sons, pp. 18–23

Hobbs, J. E., J. A. Lindesay and H. A. Bridgman (Eds) (1998) *Climates of the Southern Continents: Present, Past and Future*, Chichester, John Wiley & Sons

Hogwood, B. W. and L. A. Gunn (1984) *Policy Analysis for the Real World*, Oxford, Oxford University Press

Holmes, J. (2002) 'Diversity and change in Australia's rangelands; a post-productivist transition with a difference?', *Transactions of The Institute of British Geographers* **27**: 262–284

Holmes, J. H. (1997) 'Diversity and change in Australia's rangeland regions: translating resource values into regional benefits', *The Rangeland Journal* **19**: 3–25

Houghton, J. T. (1994) *Global Warming: the Complete Briefing*, Oxford, Lion Publishing

Houghton, J. T., Y. Ding, D. J. Griggs, M. Noguer, P. J. van der Linden and D. Xiaosu, (Eds) (2001) *Climate Change 2001: The Scientific Basis*, Cambridge, Cambridge University Press

Hughes, T. (1999) *Bush Weak*, The Walkley Magazine, August: 5

Hugo, G. (2001) 'Population centenary article, A century of population change in Australia', in Australian Bureau of Statistics (Ed.), *Year Book Australia 2001*

Hy, R. J. and W. L. Waugh, Jr (1990) 'The Function of Emergency Management', in Waugh, W. L., Jr and R. J. Hy (Eds), *Handbook of Emergency Management: Programs and Policies Dealing with Major Hazards and Disasters*, New York, Greenwood Press, pp. 11–26

International Panel on Climate Change (IPCC) (2001) *Climate Change 2001: Synthesis Report*, Cambridge, Cambridge University Press

Jennings, G. and D. Stehlik (2000) 'Agricultural women in Central Queensland and changing modes of production: a preliminary exploration of the issues', *Rural Society* **10**(1): 63–78

Karoly, D., J. Risbey and A. Reynolds (2003) 'Global warming contributes to Australia's worst drought', Sydney, WWF Australia, http://www.wwf.org.au

Keenan, S. P. and R. S. Krannich (1997) 'The social context of perceived drought vulnerability', *Rural Sociology* **62**(1): 69–88

Ker Conway, J. (1989) *The Road from Coorain – an Australian Memoir*, London, Heinemann

Kingdon, J. W. (1995) *Agendas, Alternatives, and Public Policies* (2nd Edition), New York, HarperCollins College Publishers

Knight, F. H. (1921) *Risk, Uncertainty and Profit*, New York, Century Press

Knutson, C., M. Hayes and T. Phillips (1998) *How to Reduce Drought Risk*, A guide prepared by the Preparedness and Mitigation Working Group of the Western Drought Coordination Council Lincoln, Nebraska, National Drought Mitigation Center

Krause, M. (1995) *Rural Property Planning, Risk Management*, Chatswood, Inkata Press

Lawrence, G., I. Gray, D. Stehlik and H. Bulis (1997) 'Agricultural restructuring under conditions of drought in Central Queensland', in Burch, D., G. Lawrence, R. E. Rickson and J. Goss (Eds), *Australiasian Food and farming in a globalised economy: recent developments and future prospects* Department of Geography and Environmental Sciences: Monash University No 50: 3–14

Lawrence, G., D. Stehlik and I. Gray (1999) 'Changing spaces: The effects of macro-social forces on rural Australia', in Kasimis, B., and A. Papadopoulos (Eds), *Local Responses to Global Integration. Exploring the Socio-economic Aspects of Rural Restructuring* London, Ashgate Publishing, pp. 63–87

Lawson, H. (1988) 'Beaten Back' in Baglin, D. (Ed.), *Henry Lawson's Images of Australia* Frenchs Forest, NSW, Reed Books, pp.82

Levin, S. A. (1998) 'Ecosystems and the biosphere as complex adaptive systems', *Ecosystems* **1**: 431–436

Linacre, E. T. and B. Geerts (1997) *Climates and Weather Explained*, London, Routledge

Lindblom, C. E. (1959) 'The science of "muddling through"', *Public Administration Review* **19**: 79–82

Lindblom, C. E. (1965) *The Intelligence of Democracy: Decision Making Through Mutual Adjustment*, New York, Free Press

Lindesay, J. A. and K. M. Johnson (2003) 'Changing seasonality in Australian rainfall', *Preprints of the 7th International Conference on Southern Hemisphere Meteorology and Oceanography*, Wellington, New Zealand, American Meteorological Society, Boston

Lovett, J. V. (Ed.) (1973) *The Environmental, Economic and Social Significance of Drought* Sydney, Angus & Robertson

Lowi, T. J. (1979) *The End of Liberalism : The Second Republic of the United States* (2nd Edition), New York, Norton

Malcolm, L. R. (1992) 'Farm risk management and decision making', in Trapnell, L. N., and W. W. Fisher (Eds) *Proceedings of national workshop on risk management, 9–11 November 1992*, Melbourne, Department of Natural Resources and Environment, pp. 69–86

Malcolm, L. R. (1994) 'Managing farm risk: There may be less to it than is made of it', in Powell, R. (Ed.) *Proceedings of Conference: Risk Management in Australian Agriculture*, The University of New England, Armidale, NSW, 15–16 June 1994, pp. 180–201

Martin, P., M. Lubulwa, C. Riley and S. Helali (2000) 'Farm performance, managing risks', in Australian Bureau of Agricultural and Resource Economics (Ed.) *Proceedings of the National Outlook Conference*, Canberra, Australian Bureau of Agricultural and Resource Economics, 29 February–2 March 2000

Matthews, R. A. J. (2000) 'Facts versus factions: the use and abuse of subjectivity in scientific research', in Morris, J. (Ed.), *Rethinking Risk and the Precautionary Principle*, Oxford, Butterworth-Heinemann, pp. 247–282

May, P. J. (1992) 'Policy learning and failure', *Journal of Public Policy* **12**(4): 331–354

Mayers, B. (1995) *Insurance Based Risk Management for Drought*, Occasional Paper CV02/95 Canberra, Land and Water Resources Research and Development Corporation and Rural Industries Research and Development Corporation,

McCall, M. W. and R. E. Kaplan (1990) *Whatever it Takes – The Realities of Managerial Decision Making*, New Jersey, Prentice-Hall

McCown, R. L., P. S. Carberry, M. A. Foale, Z. Hochman, J. A. Coutts and N. P. Dalgliesh (1998) 'The FARM.SC.APE approach to farming systems research', in Michalk, D. L., and J. E. Pratley (Eds) *Proceedings of the 9th Australian Agronomy Conference*, Charles Sturt University, Wagga Wagga, pp. 633–636

McKeon, G. M., S. M. Howden, N. O. J. Abel and J. M. King (1993) 'Climate change: adapting tropical and subtropical grasslands', *Proceedings of the XVIIth International Grassland Congress*, Palmerston North, New Zealand, 13–16 February 1993

Megalogenis, G. and A. Wahlquist (2002) 'Our greenest drought', *The Weekend Australian* October 26–27

Monnik, K. (2000) 'Role of drought early warning systems in South Africa's evolving drought policy', in Wilhite, D. A., M. V. K. Sivakumar and D. A. Wood (Eds), *Early Warning Systems for Drought Preparedness and Drought Management, Proceedings of an Expert Group Meeting, Lisbon, Portugal, 5–7 September* Geneva, Switzerland, World Meteorological Organization: 47–56

Moss, D. A. (2002) *When All Else Fails: Government as the Ultimate Risk Manager*, Cambridge USA, Harvard University Press

Munro, R. K. and L. M. Leslie, (Eds) (1997) *Climate Prediction for Agricultural and Resource Management*, Canberra, Bureau of Resource Sciences

Munro, R. K. and M. J. Lembit (1997) 'Managing climate variability in the national interest: Needs and objectives', in Munro, R. K., and L. M. Leslie (Eds) *Climate prediction for agricultural and resource management: Australian Academy of Science Conference*, Canberra, 6–8 May 1997

National Drought Policy Commission (2000) *Preparing for Drought in the 21st Century: Executive Summary*, Washington, DC, 2000

National Land and Water Resources Audit (NLWRA) (2002) *Australia's Natural Resources: 1997–2002 and beyond* Canberra, NLWRA

Nelson, R. A., D. P. Holzworth, G. L. Hammer and P. T. Hayman (2002) 'Infusing the use of seasonal climate forecasting into crop management practice in North East Australia using discussion support software', *Agricultural Systems* **74**: 393–414

Nicholls, N. (1997) 'The centennial drought', in Webb, E. K. (Ed.) *Windows on Meteorology–Australian perspective*, Melbourne, CSIRO Publishing, pp. 183–204

Nicholls, N. (1997) 'Developments in climatology in Australia: 1946–1996', *Australian Meteorological Magazine* **46**: 127–135

Nicholls, N. (1999) 'Cognitive illusions, heuristics and climate prediction', *Bulletin of the American Meteorological Society* **80**: 1385–97

Nicholls, N. and K. K. Wong (1990) 'Dependence of rainfall variability on mean rainfall, latitude, and the Southern Oscillation', *Journal of Climate* **3**: 163–170

Nicholls, N. and A. Henderson Sellers (1991) 'The El Nino/Southern Oscillation and Australian vegetation', *Advances in Vegetation Science* **12**: 23–36

NSW Farmers Association (2001) *Submission to the review of natural disaster relief and mitigation arrangements*

O'Meagher, B., M. Stafford-Smith and D. H. White (2000) 'Approaches to integrated drought risk management: Australia's National Drought Policy' in Wilhite, D. A. (Ed.) *Drought: A Global Assessment*, London, Routledge Publishers Volume 2, pp. 115–128

Ostrom, E. (1999) 'Coping with tragedies of the commons', *Annual Review of Political Science* **2**: 493–535

Parliament of NSW (1966) *Select Committee of the Legislative Assembly upon Drought Relief – Second Progress Report*, Sydney, Government Printer

Patton, D. (1993) 'The ABCs of risk assessment', *EPA Journal*; Environmental Protection Agency, Washington, DC January–March: 10–15

Peart, G. (1992) 'Pastures, livestock and the bottom line', in Michalk, D. (Ed.) *Proceedings of the 7th Annual Conference of the Grasslands Conference*, Tamworth, 8–9 July 1992, pp. 74–81

Powell, R. (Ed.) (1994) *Proceedings of Conference Risk Management in Australian Agriculture, 15–16 June 1994* Armidale, NSW, University of New England

Power, S. B., F. Tseitkin, V. Mehta, B. Lavery, S. Torok and N. Holbrook (1999) 'Decadal climate variability in Australia during the twentieth century', *International Journal of Climatology* **19**: 169–184

Power, S. B., T. Casey, C. Folland, A. Colman and V. Mehta (1999) 'Inter-decadal modulation of the impact of ENSO on Australia', *Climate Dynamics* **15**: 319–324

Pyne, S. (1991) *Burning Bush, A Fire History of Australia* London, University of Washington Press

Quiggin, J., G. Karagiannis and J. Stanton (1993) 'Crop insurance and crop production: An empirical study of moral hazard and adverse selection', *Australian Journal of Agricultural Economics* **37**(2): 95–113

Rabat Declaration (2001) *Meeting on Opportunities for Sustainable Investment in Rainfed Areas of West Asia and North Africa*, Rabat, Morocco, 25–26 June 2001

Ribot, J. C. (1996) 'Introduction. Climate variability, climate change and vulnerability: Moving forward by looking back', in Ribot, J. C., A. R. Magalhães and S. S. Panagides (Eds), *Climate Variability, Climate Change and Social Vulnerability in the Semi-arid Tropics*, Cambridge, Cambridge University Press

Ricks, B. (2002) 'Drought tightens grips', *The Weekend Bulletin* 20 July 2002

Robbins, P. F., N. Abel, H. Jiang, M. Mortimer, M. Mulligan, G. S. Okin, D. M. Stafford Smith and B. L. Turner, II (2002) 'Group 2: What are the key dimensions at the community scale?', in Reynolds, J. F., and D. M. Stafford Smith (Eds), *Dahlem Workshop Report 88*. Berlin, Dahlem University Press, pp. 325–355

Rosen, H. S. (1999) *Public Finance*, Boston, Irwin McGraw-Hill

Saji, N. H., B. M. Goswami, P. N. Vinayachandran and T. Yamagata (1999) 'A dipole mode in the tropical Indian Ocean', *Nature* **401**: 360–363

Saunders, D. A., A. J. M. Hopkins and R. A. How, (Eds) (1990) 'Australian Ecosystems: 200 Years of Utilization, Degradation and Reconstruction', *Proceedings of the Ecological Society of Australia*, 16

Schapper, H. P. (1970) 'Elements of a national policy for Australian agriculture', *Farm Policy* **9**(4): 91–99

Scott, B. (1994) 'Farmers' risk management attitudes, perceptions and practices', in Powell, R. (Ed.) *Proceedings of Conference: Risk Management in Australian Agriculture*, The University of New England, Armidale, NSW, 15–16 June 1994

Senate Rural and Regional Affairs and Transport References Committee (1995) *The Impact of assets tests on farming families access to Social Security and AUSTUDY: second report – Social Security assets tests* Canberra, AGPS

Senate Standing Committee on Rural and Regional Affairs (1992) *A national drought policy – appropriate government responses to the recommendations of the Drought Policy Review Task Force : final report* Canberra, The Parliament of the Commonwealth of Australia

Shakespeare, W. (1996) *King Lear* Hunter, G. K. (Ed.) London, Penguin Books

Shanteau, J. (1992) 'Decision making under risk: Applications to insurance purchasing', *Advances in Consumer Research* **19**: 177–181

Shaw, A. G. L. (1967) 'History and development of Australian agriculture', in Williams, D. B. (Ed.), *Agriculture in the Australian Economy*, Sydney, Sydney University Press

Simmons, P. (1993) 'Recent developments in Commonwealth drought policy', *Review of Marketing and Agricultural Economics* **61**(3): 443–454

Smith, D. I. (1993) 'Drought policy and sustainability: Lessons from South Africa', *Search* **24**(10): 292–295

Smith, D. I., M. F. Hutchinson and R. J. McArthur (1992) *Climatic and Agricultural Drought: Payments and Policy*, Canberra, Australian National University

Snedden, the Hon Billy, QC MP (1971) *Commonwealth Payments to or for the States 1971–72*, Parliamentary Paper No 54 Canberra, Commonwealth Government Printing Service, 17 August 1971

Spencer, B. (Ed.) (1896) *Report on the work of the Horn Scientific Expedition to central Australia. Part I. Introduction, narrative, summary of results, supplement to zoological report, map* Bundaberg, Queensland, Facsimile (1994) – Corkwood Press

Stafford Smith, D. M. (1994) 'Sustainable production systems and natural resource management in the Rangelands', *Proceedings ABARE Outlook Conference*, Canberra, ABARE, February 1994

Stafford Smith, D. M. and B. D. Foran (1990) 'RANGEPACK: the philosophy underlying the development of a microcomputer-based decision support system for pastoral land management', *Journal of Biogeography* **17**: 541–546

Stafford Smith, D. M. and S. R. Morton (1990) 'A framework for the ecology of arid Australia', *Journal of Arid Environments* **18**: 225–278

Stafford Smith, D. M., J. F. Clewett, A. M. Moore, G. M. McKeon and R. Clark (1997) *DroughtPlan. Full Project Report.* DroughtPlan Working Paper No 10 Canberra, CSIRO Alice Springs/LWRRDC Occasional Paper Series

Stafford Smith, M. and G. M. McKeon (1996) *Assessing the historical frequency of drought events on rangelands grazing properties: case studies* Canberra, Bureau of Rural Sciences

Stafford Smith, M. and G. M. McKeon (1998) 'Assessing the historical frequency of drought events on grazing properties in Australian rangelands', *Agricultural Systems* **57**: 271–299

Stafford Smith, M., S. R. Morton and A. J. Ash (2000) 'Towards sustainable pastoralism in Australia's rangelands', *Australian Journal of Environmental Management* **7**(4): 190–203

Stafford Smith, M., J. Cross and J. Breen (2001) *Taxation instruments and grazing enterprises: RISKHerd regional reports* RISKHerd Project Report No. 6 Alice Springs, CSIRO

Stafford Smith, M. and J. F. Reynolds (2002) 'Desertification: A new paradigm for an old problem', in Reynolds, J. F., and D. M. Stafford Smith (Eds), *Dahlem Workshop Report 88* Berlin, Dahlem University Press, pp. 403–425

Stehlik, D., I. Gray and G. Lawrence (1999) *Drought in the 1990s. Australian Farm Families' Experiences* Canberra, Rural Industries Research and Development Corporation 99/14. UCQ-5A

Stehlik, D., G. Lawrence and I. Gray (2000) 'Gender and drought: Experiences of Australian women in the drought of the 1990s', *Disasters* **24**(1): 38–53

Stehlik, D. and L. Chenoweth (2003 forthcoming) *Integrating Social Capital with Resiliency: A Transformative and Innovative Community Building (TICB) model*, mimeo

Stewart, M., G. Reid, S. Jackson, L. Buckles, W. Edgar, C. Mangham and N. Tilley (1996/97) 'Community resilience: Strengths and challenges', *Health and Canadian Society. Special Edition: Resiliency* **4**(1): 53–82

Stiglitz, J. E. (1988) *Economics of the Public Sector*, New York, W W Norton & Co

Sturman, A. P. and N. Tapper (1996) *The Weather and Climate of Australia and New Zealand*, Melbourne, Oxford University Press

Tadesse, T. (2000) 'Drought and its predictability in Ethiopia', in Wilhite, D. A. (Ed.) *Drought: A Global Assessment*, London, Routledge Publishers Volume 1, pp. 135–142

The Australian (2002a) 'Farmers need their stockpile of good luck', 13 June 2002

The Australian (2002b) 'Lending hand to farmers benefits us all', 3 October 2002

Thompson, D., D. Jackson, N. Tapp, N. Milham, R. Powell, B. Douglas, G. Kennedy, E. Jackman and G. White (1996) *Analysing Drought Strategies to Enhance Farm Financial Viability*, Armidale, University of New England

Thompson, D. and R. Powell (1998) 'Exceptional circumstances provisions in Australia – is there too much emphasis on drought', *Agricultural Systems* **57**(3): 469–488

Trapnell, L. N. and W. W. Fisher (Eds) (1993) *Incorporating Risk into Decision Support Systems and Farm Business Management Systems: Proceedings of national workshop, 9–11 November 1992*, Melbourne, Department of Natural Resources and Environment

Truss, the Hon Warren, MP (2002a) 'Early help for drought-affected northwest NSW farmers', Media Release by the Federal Minister for Agriculture, Fisheries and Forestry, 19 September 2002

Truss, the Hon Warren, MP (2002b) 'Commonwealth to push States/Territories to put drought-stricken farmers first', Media Release by the Federal Minister for Agriculture, Fisheries and Forestry, 2 October 2002

Ubergang, J. W. (2002) 'The Crooble Plan: To reduce the vulnerability of Australia to weather and climate extreme disasters', *North West Magazine*, December 2: 6

UNCCD (1999) *United Nations Convention to Combat Desertification (text with annexes)*, Bonn, Germany

UNDP/UNSO (2000) *Report on the Status of Drought Preparedness and Mitigation in Sub-Saharan Africa*, New York, UN Development Program, Office to Combat Desertification and Drought

Van Manen, M. (1990) *Researching Lived Experience. Human Science for an Action Sensitive Pedagogy*, New York, State University of New York Press

Vanclay, F., L. Mesiti and P. Howden (1998) 'Styles of farming and farming subcultures: Appropriate concepts for Australian rural sociology?', *Rural Society* **8**(2): 85–107

Wahlquist, Å. (1991a) 'A drought to wither all hope', *Sydney Morning Herald*, 20 May 1991

Wahlquist, Å. (1991b) 'Farmers trapped on the land', *Sydney Morning Herald*, 6 June 1991

Wahlquist, Å. (1994a) 'Appeal near end after $124 m', *Sydney Morning Herald*, 22 October 1994

Wahlquist, Å. (1994b) 'One man's recipe for surviving the big dry', *Sydney Morning Herald*, 7 December 1994

Wahlquist, Å. (1995) 'Farming our parched land', *Sydney Morning Herald*, 9 April 1995

Wahlquist, Å. (1996) 'Harder rains are yet to fall', *The Bulletin*, 9 January 1996

Wahlquist, Å. (1999) 'Accentuate the positive, Fischer tells farmers', *The Weekend Australian*, 23–24 January 1999

Wahlquist, Å. (2000) 'Reporting of Rural Australia: fearless or fair?', *Australia Centre for Independent Journalism Seminar*, Armidale, 9 October 2000

Wahlquist, Å. and M. Kidman (1994) 'Drought now worst in history', *Sydney Morning Herald*, 31 December 1994

Walker, B., S. Carpenter, J. Anderies, N. Abel, G. Cumming, M. Janssen, L. Lebel, J. Norberg, G. A. Peterson and R. Pritchard (2002) 'Resilience management in social-ecological systems: a working hypothesis for a participatory approach', *Conservation Ecology* **6** (1): 14. [online] http://www.consecol.org/vol6/iss1/art14

Walker, B. H. and M. A. Janssen (2002) 'Rangelands, pastoralists and governments: interlinked systems of people and nature', *Philosophical Transactions of the Royal Society of London* **357**: 719–725

Walsh, P. (1994) 'Cassandra', *Australian Financial Review*, 18 July 1994

Webb, P., J. von Braun and Y. Yohannes (1992) *Famine in Ethiopia: policy implications of coping failure at national and household levels*, International Food Policy Research Institute Research Report No 92 Washington DC

West, B. and P. Smith (1996) 'Drought, discourse and Durkheim: a research note', *Australian and New Zealand Journal of Sociology* **32**(1): 93–102

Whetton, P. H., J. J. Katzfey, K. J. Hennessy, X. Wu, J. L. McGregor and K. C. Nguyen (2001) 'Developing scenarios for climate change for Southeastern Australia: an example using regional climate model output', *Climate Research* **16**: 181–201

White, B. J. (2000) 'The importance of climate variability and seasonal forecasting to the Australian economy', in Hammer, G. L., N. Nicholls and C. Mitchell (Eds), *Applications of Seasonal Climate Forecasting in Agricultural and Natural Ecosystems – the Australian Experience*, The Netherlands, Kluwer Academic, pp. 1–22

White, D., D. Collins and M. Howden (1997) 'Drought in Australia: prediction, monitoring, management and policy', in Wilhite, D. A. (Ed.), *Drought Assessment: Management and Planning*, Boston, Kluwer, pp. 213–236

White, D. H. (2000) 'Drought policy, monitoring and management in arid lands', *Annals of the Arid Zone* **39**: 105–129

White, D. H., S. M. Howden, J. J. Walcott and R. M. Cannon (1995) 'Estimating the extent and variability of drought', in Binning, P., H. Bridgman and B. Williams (Eds) *International Congress on Modelling and Simulation Proceedings. MODSIM. Vol. 2: Air Pollution and Climate*, The University of Newcastle: 255–259

White, D. H. and L. Karssies (1997) 'Australia's National Drought Policy: Aims, analyses and implementation', *IXth World Water Congress*, Montreal, Canada, 1–6 September 1997

White, D. H., G. Tupper and H. S. Mavi (1999) *Agricultural climate research and services in Australia* Land and Water Resources Research and Development Corporation LWRRDC Occasional Paper CV02/99 Canberra

White, D. H., D. A. Wilhite, B. O'Meagher and G. L. Hammer (2001) 'Highlights of drought policy and related science in Australia and the USA', *Water International* **26**(3): 349–357

White, W. B. (2000) 'Influence of the Antarctic Circumpolar Wave on Australian precipitation from 1958 to 1997', *Journal of Climate* **13**: 2125–2141

Whitmore, J. S. (2000) *Drought Management of Farmland*, Dordrecht, Kluwer Academic Publishers

Wildavsky, A. B. (1979) *Speaking truth to power: the art and craft of policy analysis*, Boston, Little Brown

Wilhite, D. A. (1991) 'Drought planning: A process for State Government', *Water Resources Bulletin* **27**(1): 29–38

Wilhite, D. A. (1993) 'Planning for drought: A methodology', in Wilhite, D. A. (Ed.), *Drought Assessment, Management and Planning: Theory and Case Studies*, Norwell, Kluwer Academic Publishers, pp. 87–108

Wilhite, D. A. (1994) 'State-level drought planning in the United States: Factors influencing plan development', *Water International* **19**: 15–24

Wilhite, D. A. (1997) 'State actions to mitigate drought: Lessons learned', *Journal of the American Water Resources Association* **33**(5): 961–68

Wilhite, D. A. (Ed.) (2000a) *Drought: A Global Assessment*, London, Routledge Publishers

Wilhite, D. A. (2000b) 'Drought as a natural hazard: Concepts and definitions', in Wilhite, D. A. (Ed.), *Drought: A Global Assessment* London, Routledge Volume 1, pp. 3–18

Wilhite, D. A. (2001) 'Moving beyond crisis management', *Forum for Applied Research and Public Policy* **16**(1): 20–28

Wilhite, D. A. and M. H. Glantz (1985) 'Understanding the drought phenomenon: The role of definitions', *Water International* **10**: 111–120

Wilhite, D. A. and M. H. Glantz (1987) 'Understanding the drought phenomenon: the role of definitions', in Wilhite, D. A., W. E. Easterling and D. A. Wood (Eds), *Planning for Drought: Toward a Reduction of Societal Vulnerability* Boulder, Colorado, Westview Press, pp. 11–30

Wilhite, D. A. and O. Vanyarkho (2000a) 'Drought: Pervasive impacts of a creeping phenomenon', in Wilhite, D. A. (Ed.), *Drought: A Global Assessment* London, Routledge Publishers Volume 1, pp. 245–255

Wilhite, D. A., M. V. K. Sivakumar and D. A. Wood (2000b) 'Improving drought early warning systems in the context of drought preparedness and mitigation, summary of breakout sessions', in Wilhite, D. A., M. V. K. Sivakumar and D. A. Wood (Eds), *Early Warning Systems for Drought Preparedness and Drought Management, Proceedings of an Expert Group Meeting, Lisbon, Portugal, 5–7 September* Geneva, Switzerland, World Meteorological Organization

Wilhite, D. A., M. J. Hayes, C. Knutson and K. H. Smith (2000) 'Planning for drought: moving from crisis to risk management', *Journal of American Water Resources Association* **36**: 697–710

Index